MOUNTAIN BUILDING

MOUNTAIN BUILDING

A study primarily based on Indonesia
region of the world's most active
crustal deformations

by

REIN W. VAN BEMMELEN

University of Utrecht

with an introduction by

RAYMOND C. MOORE

State University of Kansas

Springer-Science+Business Media, B.V.

1954

Part II translated from the Dutch ("De Geologische Geschiedenis van Indonesië", Van Stockum, The Hague, 1952) by JAN HOSPERS, PH. D., F.G.S.

ISBN 978-94-017-5730-0   ISBN 978-94-017-6087-4 (eBook)
DOI 10.1007/978-94-017-6087-4

INTRODUCTION

BY RAYMOND C. MOORE *

The subject of mountain building is one of the most interesting and important of all geological studies, for crumpling of the earth's lithosphere in many linear tracts has operated throughout the history of our planet to accentuate topographic relief of the crust. Thus elevation of land masses above sea level has been maintained and processes of denudation and sedimentation perpetuated; otherwise all continents and islands would have vanished millions of years ago beneath a universal ocean. The lands have been rejuvenated by uplifts which culminate in mountain chains. Evidence from historical geology shows that these uplifts have occurred periodically rather than continuously and that successive epochs of mountain-building deformations are separated from one another by relatively long periods of quiescence. The placement of mountain systems, past and present, is not at random but it shows certain relationship to rigid and mobile sectors of the crust. Both the time and place of mountain building are correlated with evolution of the continental and oceanic areas of the globe.

What are the explanations of these things? Answers to questions concerning the causes of mountain building and the mode of making elevated mountain chains are in part to be sought by examining characters of mountain distribution in space and time during the course of earth history. Also, features of mountain structure, including the nature of folding and faulting of rock strata, the occurrence

---

* Professor of Geology and Head of the Department of Geology, University of Kansas; also State Geologist of Kansas and Director of Research, State Geological Survey of Kansas; in 1951–52, Visiting Fullbright Professor at the Rijksuniversiteit te Utrecht, chosen as representative of the Association of American Universities.

of metamorphosed rocks, and the character of associated igneous rock bodies, should supply significant evidence. Observations of the distribution of gravity, computations of departure from isostatic equilibrium, analyses of data pertaining to earthquakes and volcanism within or adjacent to mountain belts, and all obtainable information concerning physicochemical conditions within and beneath the earth crust are likely to have important bearing on recognition of what are the controlling factors in orogeny. Both the sources of energy and the mechanism of their operation must be defined in studying the principles of mountain building.

The fitness of any geologist to formulate a coherent synthesis of the complexly interrelated chemical, physical, and temporal factors involved in mountain building depends on two chief requirements. These are: (1) personal attributes which especially include possession of a keen intellect coupled with thorough training in fields that include chemistry, physics, mathematics, and geology; skill in discernment between the relevant and irrelevant; and the gift of a reasonably wellcontrolled scientific imagination; and (2) the opportunity to study one or more exceptionally clear examples of mountain building which furnish a maximum range of unobliterated evidence concerning successive stages in evolution. In my opinion, the author of this book, Prof. Dr. Ir. Rein W. van Bemmelen has such fitness. His thorough geological and engineering training has been employed chiefly during many years in studies of mountain structures in various parts of the world where significant relationships in geophysical setting and geological history are revealed. Especially, he has had opportunity to survey extensively the structural evolution of the Indonesian region.

Advancement of scientific knowledge certainly is accelerated and made more sure by finding the best available materials and methods for any investigation. Thus, the fruit fly *Drosophila* has been discovered to offer supreme research materials for studies in the field of genetics, serving to define clearly relations between chromosome structure and body characters that would be very difficult to establish by work on other organisms. In stratigraphy, the fossiliferous Jurassic deposits of central and northwestern Europe have first-rank importance because they have furnished basis for fundamental concepts of biostratigraphic zonation and correlation; also they first served to demonstrate characters of contemporaneous lateral variations in sedimentation, introducing concepts of facies. Late Paleozoic deposits of the central United States provide surpassingly clear demonstration of cyclic sedimentation involving many alternations of continental and neritic marine deposits; with little doubt, the best laboratory in the world for learning principles of such cyclic sedimentation is found in Kansas and adjoining territory.

In exactly similar manner, the Indonesian area offers the best opportunity in the world for elucidation of various phenomena associated with mountain building. It is a region in which interrelated volcanic activity, seismicity, belts of strongly defined gravity anomalies, and chemically distinct suites of igneous rocks are correlated with parallel zones of downwarped crust (geosynclines) and more or less strongly elevated crust (geanticlines), the latter including mountain belts such as the "backbones" of Sumatra, Java, and Borneo. Mountain building is in progress at the present moment in Indonesia, and the evidence organized by van Bemmelen shows that in this region similar mountain building has proceeded in pulsatory manner and laterally shifting placement throughout the last 200 million years or more of earth history. Certainly, the East Indies provide exceptional opportunity for delineation of the principles of mountain building. Accordingly, the concisely written description of Indonesian geology by van Bemmelen presented in this volume has worldwide interest, because conclusions derived from it have application to all parts of the globe and all of geologic history. It is not an essay on regional geology having value only for those who want to know something of the geological structure and history of a portion of Australasia.

The writing of this introduction was requested by Professor van Bemmelen in spite of my suggestion that it seems superfluous. In small degree I contributed to shaping the present book by recommending modification of its originally chosen title and incorporation in the text of a general statement concerning principles of mountain building (Part I), to be followed by description of the evolution of mountain belts and associated geological phenomena in the Indonesian region (Part II). This makes more evident the far-reaching importance of territory in the East Indies for study of mountain building generally. Part II consists of a translation of the author's "De geologische geschiedenis van Indonesië" (Geological history of Indonesia), which was published in 1952 while I was guest member of the faculty of the Mineralogisch-geologisch Instituut of the Rijksuniversiteit te Utrecht, and thus as a colleague of Professor van Bemmelen had opportunity to discuss with him the desirability of making the contents of this book available to English-speaking geologists.

The name of van Bemmelen is very well known to Dutch, English and other European geologists but not to most Americans. His comprehensive 2-volume work on the "Geology of Indonesia", published in 1949, firmly establishes his reputation as an authority on this region. In addition, Professor van Bemmelen has done much field work in mountain and volcanic areas of Spain, Scotland, France Austria, Switzerland, Belgium, Germany, and Iceland. He has published numerous papers, particularly on geophysical and geochemical aspects of mountain building, although most of these in Dutch and other European journals are virtually unknown in North America. Having noted these things, I shall venture to offer comments on Professor van Bemmelen's presentation of principles of mountain building.

The hypotheses proposed by various workers to account for mountain building are classified by van Bemmelen in two groups. In what he terms the "unicausalistic-mobilistic" group, he clearly emphasizes the underlying concepts of crustal yielding to laterally exerted compressional forces. In what he terms the "bicausalistic-fixistic" group, crustal stability in over-all geographic position accompanies dominantly vertical movements and gravity-controlled secondary adjustments. The movements of geanticlinal elevation and geosynclinal depression are inferred to be associated with geochemical and geophysical changes in the crust that involve differentiation of more basic and more acid fractions, and plutonic activity is conceived to be the primary cause of crustal disturbance. Parenthetically, one may observe that van Bemmelen employs the words "hypothesis" and "theory" quite interchangeably, ignoring the distinction that assigns to "theory" an intermediate status between "hypothesis" and "law"; out of many hypotheses, adequate testing may produce a theory, and ultimately this may receive acceptance as scientific law. It is emphatically my opinion that no hypothesis of mountin building is entitled to be ranked as a theory.

Turning to nomenclature of the hypothesis groups defined by

van Bemmelen, most American geologists will be repelled by such terms as "unicausalistic-mobilistic" and „bicausalistic-fixistic", even though they understand their meaning and concur in the ideas. Actually, argument can be advanced that more than a single cause leads to crustal compression, and also that successive parts of the "chain reaction" postulated and described by van Bemmelen in elucidating "bicausalistic-fixistic" concepts can be construed either as having a single initiating cause or being divisible into several measurably independent although interlocked causes. Likewise, special meanings must be assigned to "mobilistic" and "fixistic" in order to make them truly acceptable. According to dominant influences adduced in differentiating these two groups of hypotheses, it seems to me sufficient and more simple to class the first ("unicausalistic-mobilistic") as lateral compression hypotheses and the second ("bicausalistic-fixistic") as plutonic hypotheses. *

Surely, it is inappropriate here to undertake an analysis of the hypothesis of mountain building which van Bemmelen terms the "undation theory". This is actuated primarily by plutonic forces and characterized in later evolution by gravity thrusts that may have very appreciable lateral components. Attention may be called rather, to the admirable synthesis of pertinent geological observations which the author of this book has made in drawing on his knowledge of the geomorphology, stratigraphy, structure, geophysical state, igneous geology (including volcanism), seismicity, and geological history of Indonesia. An objective consideration of the evidence which he has brought together can hardly fail to yield agreement with several important conclusions stated, such as the significance of placement of ophiolites and other suites of igneous rocks, relationships of geanticlines and geosynclines to gravity anomalies, meaning of volcanic and nonvolcanic arcs, control of the locus and nature of sedimentation; and especially the systematic lateral shift of orogenic belts in geological history of this region. Collectively, this evidence is compelling, even though it does not establish the nature of changes in the lithosphere extending downward some 40 km. as plausibly postulated by van Bemmelen. Such

---

* The author agrees with MOORE's suggestion to call the group of unicausalistic-mobilistic theories "lateral compression hypotheses", and the bicausalistic-fixistic ones "plutonic hypotheses", on the condition that these terms are understood as defined in Part I.

postulates are very legitimate hypotheses, and remembering that they are neither more nor less than hypotheses, credence in them is merited in such degree as agreement with all possible tests is found. Viewed in this way, it is highly desirable that the author's picture of interrelated causes and processes controlling the evolution of mountains should be very widely known to geologists. Although some doubtless will find it difficult, if not impossible, with open mind to contemplate concepts which to them are new and strange, the idea that essentially localized plutonic forces give rise to crustal relief sufficient to generate pronounced crumpling, faulting, and metamorphism of rocks by gravitative readjustments needs to be analyzed carefully and judicially. Given requisite geological time and appropriate physicochemical conditions, mountain-forming disturbance of the earth crust in a given region very conceivably involve such large "epidermal", "dermal", and "bathydermal" deformations under the combined effects of plutonic changes and gravitative forces as are advocated in the mountain-building hypothesis of van Bemmelen.

This book is recommended for thoughtful study of geologists everywhere.

Lawrence, Kansas   RAYMOND C. MOORE
15 September 1953

x

CONTENTS

PART I. PRINCIPLES OF MOUNTAIN BUILDING

PART II. THE OROGENIC EVOLUTION OF THE EARTH'S CRUST IN INDONESIA

I. INTRODUCTION

II. STRATIGRAPHY

III. VOLCANISM

PRINCIPLES OF MOUNTAIN BUILDING

Man and Mountains

Why are we interested in mountain building? Is it one of those very specialized topics of scientific research, which presents only remote relations with our daily struggle for life? Or has it a more direct meaning for our actions?

The surface of our planet is the homestead of Man. It provides the minerals and part of the energy for our agricultural and industrial production; its relief influences the climates and the routes of trade. The land under our feet is the most important part of our surroundings, for it is the natural basis of the biological "niche" in which we live. If we want to understand something of the possibilities of human life, and if we want to push back the limitations imposed on us by our environment, then we should investigate the formation of that niche. Its basis is the earth's crust. How did that base of our life come into existence, how did its mineral composition originate, how was its relief created? These problems are studied by Geology.

The geologist distinguishes internal and external sources of energy. The endogenic energy causes the structural deformations and chemical transformations of the crust, with accompanying phenomena of volcanism, earthquakes and deviations from isostatic or floating equilibrium of the crust. It creates the primary crustal relief and determines the crustal composition.

The external sources of energy, derived from the sun's radiation and the tides, tend to break down the primary relief. These forces carve the valleys and push back the coasts. If the endogenic energy were no more active, all crustal relief would disappear below sealevel and human life would become impossible.

1

The structural deformations and chemical transformations of the crust also determine the location of our mines, quarries and drillings for the supply of our mineral raw materials.

Therefore, it is natural for this age of scientific research and industrial development that there is a general interest in the problem of mountain building. *Mountain building provides the very basis of our existence on earth.*

The method of geological description

In the long and difficult process of acquiring knowledge about the relations between Mind and Nature — that is the evolution of Natural Sciences — there must always be an interaction between *inductive* remoulding of primary observations into more general rules and *deductive* interpretation of elementary facts, according to such rules. But general rules or natural laws are only working hypotheses for the description of natural relations. Logical deductions from these laws lead to predictions (prognoses) about such natural relations as can be checked by diagnostic observations. If the latter confirm the prognosis, the concerning law has found practical support; it then appears to be a successful working hypothesis. We go on applying it in our descriptions of natural relations until its logical consequences come into conflict with our observations. Then, reinterpretation of the primary facts, and modification or even complete rejection of the natural law concerned will be necessary.

However there is always the danger that our observations are no longer primary data objectively gathered, but are unconsciously influenced by theoretical conceptions. The latter may assume a dogmatic character, being no longer mere working hypotheses providing the guiding principles of our research; they may become rigid axioms, assuming in our mental processes the same value as the facts of observation. In such a situation some scientists are apt to be prejudiced in their observations by theory, which means frustration of research. Then, if discrepancies between theoretical prognoses and observational diagnoses become apparent, it is often not the theory that is revised; instead, the observations are sometimes dogmatically reinterpreted and incorrect values may be assigned to them.

It is a dangerous condition of scientific research if ,,schools'' and their masters become more important than the free voice of

2

Nature itself. Such situations occur more often than is realized by most of us, in many fields of natural science, geology not excluded.

The main purpose of this book is to present an attempt for an undogmatic approach to the principles of mountain building (Part I), and to illustrate the suggested solution of the concerning problems by the geological evolution of the Indonesian island arcs (Part II).

The conception of tangential compression

In geology, for example, many geologists, when studying a geanticlinal mountain range consisting of folded and overthrust strata, speak of the "upfolding" of such a structure as the result of tangential compressive forces, without realising that in making this statement they introduce a hypothesis. In stating the problem of mountain building it is said, for instance, that a typical orogen results from intense compression of a geosynclinal zone, in which long-continued subsidence has allowed the accumulation of great thicknesses of sedimentary rocks (LEES, 1952). However, the geological analysis of mountain ranges shows that the folding and overthrusting of the sedimentary strata often occurred at the end of the geosynclinal stage of subsidence and before the orogenic uplift. Then, after the folding and thrusting of the strata their elevation to a mountain range took place. This uplift is often accompanied by the emplacement of granite batholiths, and other igneous phenomena. Compared with the folding and overthrusting of the rock-strata the rise of a mountain range is a phenomenon of larger dimensions and of different character. So the statement that this uplift had the same cause as the folding, viz. tangential compression, is only a hypothesis. Most probably one utters, unconsciously, an unwarranted hypothesis when speaking of the "upfolding" of a mountain range.

Geologists, in studying the geometry of tectonic structures, are always impressed by the phenomena of compression, because these are preserved in folds, overthrusts and the like. However, the complementary phenomena of extension or removal of matter are less evident, because material which was in certain positions has disappeared. Little or no traces are left behind in the structure of the rock-strata. Therefore, *the importance of compression phenomena has been generally overestimated*; whereas the aggregate amount of *extension* phenomena has been systematically *underestimated*.

H. CLOOS of Bonn was one of the few geologists who clearly

3

distinguished in his tectonic descriptions the effect of extension phenomena. His experiments with clay were very instructive in this respect (CLOOS, 1936). In the Cloos-memorial volume of the Geologische Rundschau (Vol. 41, 1953) some of his pupils published papers on micro-tectonics, which elaborate this point. For instance, the observations of H. SCHOLZ on extension phenomena in the Verrucano of the Glarner nappe (Swiss Alps), and the remarks of S. KIENOW on the reduction in thickness (thinning) of the extended strata. The present author is of the opinion that the recognition of extension phenomena in rock complexes and the thinning of strata, due to stretching is a very fertile conception in modern tectonic analyses, namely in our attempts to reconstruct the mechanics of tectonic deformation from observations on rigid and consolidated rock structures.

Erosion may remove material in a fragmented or dissolved state from elevated parts of the earth's crust. It is difficult to estimate how much material disappeared by such *erosional denudation*.

But this is not the only way by which matter can be removed from relatively higher regions.

The force of gravitation may also remove material from elevated regions in a more concentrated form, like the removal of firn by glaciers, creep on mountain slopes, and slumping. These phenomena merge into other phenomena of gravitational flow, such as thinning and stretching of strata, slow spreading of mountains like "giants on feet of clay", and the like. In such cases we may speak of *tectonic denudation*.

One of the main intellectual tasks in the Natural Sciences in general, and especially in geology with its exceedingly slow and long enduring processes, is to maintain a sense of proportions. For instance, what does the factor time mean in relation to the factor strength of rocks. The strength determinations have been made in laboratories by experiments of short duration on selected, flawless samples. But the technical experimental conditions differ from those experiments made by our natural laboratory, the earth (conditions of temperature, stress, chemical environment, and local anisotropies). Calculations about the influence of the reduction of scale in model experiments on the relations between factors of viscosity and time, made by KING HUBBERT (1937), MAILLET and PAVANS DE CECCATY, and others, have shown that for the formation of mountain structures on a reduced, laboratory scale, material like "diluted tooth paste" should be used. Such material is incapable of transmitting tangential compressive forces, but it will flow under the sole action of gravity.

Gravitational flow of material (French "écoulement par gravité") becomes more and more a generally recognized form of tectonic denudation, whereby compression of strata at the foot of rising mountain chains is compensated by their thinning or removal in the higher elevated parts. Apart from this principle of superficial gliding, which is generally restricted to the sedimental epidermis, deeper (dermal and bathydermal) crustal levels can also be subjected to gravitational flow. This principle will be discussed in the chapter on tectonics in Part II (see also fig. 21).

Here we only want to point out that the phenomena compensatory to compression, namely those of spreading and tectonic denudation, are often overlooked in the field work and regional geological accounts of geologists.

General compression or local mutual compensation of compression and extension?

ED. SUESS (1886) in his masterly synthesis of our geological knowledge of about 70 years ago, adhered to the contraction theory. According to this theory, loss of volume due to cooling of our planet caused a shortening of the earth's circumference. Tangential compressive forces would be the secondary effect of this contraction, and there would be no room for extension of the crust. The formation of geanticlines and geosynclines, as well as the generation of folds and nappe-structures, would be uniformly the result of tangential compressive forces. This geologic hypothesis might therefore be called the unicausal conception of mountain building.

A serious objection against the classic contraction theory is that it is by no means certain that the earth is still cooling in post-Algonkian times. Measurements of heat flow indicate an approximate balance between the rate of cooling and the rate at which heat is generated. However, contraction is still accepted in the majority of later theories on orogenesis. Recently, variants of the contraction theory have also been proposed by L. KOBER ("Atom geology", 1951), UMBGROVE (1952), and K. K. LANDES (1952). In other geotectonic theories, the tangential compressive forces are not seen as the effect of a shortening of the earth's diameter, but are considered to be the result of thermal convection currents in the substratum, which exert a viscous drag on the overlying crust. This is the basic principle of the theories of AMPFERER (1906), SCHWINNER (1920), JOLY (1924), HOLMES (1930), VENING MEINESZ (1930), R. STAUB (1928) and others.

According to the latter theories the phenomena of compression occurring in orogenic belts are compensated by phenomena of extension and stretching in the rear of the large crustal shields of fore- or hinterland. For instance, the compression phenomena in the Alpine belt of Europe would only occur at the southern side of the African continent (R. STAUB: Der Bewegungsmechanismus der Erde, 1928). For the mobile belt itself, in which mountain building prevails, the orogenic processes would be entirely governed by tangential compression. Therefore, also according to these theories the unicausal conception is applied to the entire geosynclinal area from which an alpine-type mountain system originates, for instance the entire Tethys sea, which measures more than 1000 kilometres across.

When studying the formation of mountain ranges and foredeeps, geanticlines and geosynclines, geologists of the unicausal school explain these wave-like deformations of the earth's crust as the effect of crustal compression. Also in those cases, where the uplift is the result of isostatic rise of a mountain root, the formation of such a root is thought to be the result of crustal compression. However, the primary observations indicate only differential vertical movements of adjacent crustal tracts and zones. It is, therefore, more objective to leave this hypothesis of unicausalism out of our geological studies. At any rate, we have to consider carefully, in each case of uplift, whether it is the result of upwarping by lateral compression or merely an uplift by vertical pressure from below.

This unicausal conception is opposed by the bicausal interpretation of mountain building.

The bicausal concept of mountain building makes a distinction between differential vertical movements of neighbouring crustal parts on the one side, and differential lateral displacements (with a major horizontal component) of crustal rocks and sediments on the other. The vertical movements would cause the orogenic relief and are called primary tectogenesis. The sideward displace-

6

ments would be reactions to the vertical movements. They cause folding and thrusting of rock strata in depressed areas, compensated by tectonic denudation and thinning by spreading of the strata in adjacent elevated regions. Such gravitational reactions are called secondary tectogenesis (HAARMANN, 1930; VAN BEMMELEN, 1931).

Such mutually compensatory phenomena of compression and extension are of a much smaller regionality than those conceived by the school of unicausalism. They occur on the scale of simple anti-clines and synclines (like the collapse structures, described by HARRISON and FALCON, 1934) or on the scale of geanticlines and geosynclines of some dozens of kilometres width, like the alpine ranges of Europe or the island arcs and their side deeps in Indonesia.

The unicausal and bicausal concepts of mountain building represent two important trends in contemporary geologic thought. None of the two should assume a dogmatic value and it is necessary to discuss afresh in each individual case, whether the basic facts of regional geology fit better into the one or into the other scheme of interpretation.

Continental drift or continental growth?

The discovery of nappe-structures in the Alps and other mountain ranges of the world suggested crustal shortenings larger than could be explained by the old contraction theory. The theories of *continental* or *crustal drift*, proposed by TAYLOR (1910) and WEGENER (1915), seemed to provide a satisfactory solution.

The shortening of the crust in folded mountain belts would be compensated by lateral shifts of the adjacent crustal shields.

Two main sources of energy have been suggested for such crustal displacements: (1) the rotational energy of the earth causing drift from the poles toward the equator, and westward drift, and (2) thermal convection currents in the substratum. The first solution is considered to be inadequate by most contemporary geophysicists, but the theory of thermal convection currents is now very much in fashion.

Theories of continental drift start their account of crustal evolution by postulating a primary sialic crust which broke up into large blocks, which drifted apart (like WEGENER's „Pangea") or which extended by gravitational spreading (GUTENBERG, 1930).

Between sialic shields, or at their borders, the crustal matter and younger sediments are "upfolded" to mountain chains. ARGAND has called the congeries of such conceptions the "school of mobilism".

In VAN WATERSCHOOT VAN DER GRACHT's symposium on this new theory of continental drift (1928), we find, beside some words of favourable appreciation, severe criticisms formulated by BAILEY WILLIS, BOWIE, CHAMBERLIN, CHESTER LONGWELL, and especially SCHUCHERT. Nevertheless, the theory of continental drift is, in our day, still a widely accepted geotectonic concept. In a recent publication, for instance, R. STAUB speaks of the general advance of the African continent toward the European foreland (1952, p. 32).

On the other hand, SUESS has introduced the conception of continental growth. According to this theory, the continental nuclei occupy fixed positions at the surface of the geoid, but grow in the course of geologic history. This idea has also been put forward by M. BERTRAND (1897) almost at the same time. Adherents to theories of continental growth, which might be called the *school of fixism*, include STILLE (1924), BORN (1932) and the present author (1932, 1933). Further elaboration of the idea that continental shields do not drift, but stay where they' are, led to the conception that not only growth, but also decay of continents takes place. STILLE (1935, 1948) distinguishes primeval oceans (such as the central part of the Pacific) and oceans formed in post-Algonkian times, which originated by the breaking down and subsidence of extensive parts of the sialic crust (such as parts of the Atlantic and the Indian Ocean). Such transformations of continental shields into oceanic basins seem to be irreversible processes in the geological evolution. Moreover, parts of continents can be transformed into geosynclinal basins, and the latter may again consolidate into rigid continental blocks. This is a reversible process, as appears from the study of the Hercynian, Alpine, and other post-Algonkian mountain systems. In the second part of this book it will be pointed out that in the Indonesian region also a pre-Devonian crystalline basement complex was present, formed and consolidated by older cycles of mountain building. This primeval Indonesian continent began to subside in the course of the Paleozoic, assuming a geosynclinal character. Subsequent cycles of orogenesis transformed this geosynclinal area partly into new rigid continental blocks, such as the Sunda continent, whilst in other parts the process of mountain building still is in full swing.

According to fixistic conceptions the pattern of distribution of land and sea is primarily the effect of differential vertical movements of the sialic crust.

The energy problem (thermal or chemical?)

Theories on mountain building are ultimately related to one central problem, that of the source of endogenic energy. What kept internal geological processes going for about 3×10^9 years? Is this geological evolution a thermal or a chemical process, or both?

Most geotectonic and geophysical theories consider thermal processes (cooling of the earth, or local heating by radioactivity) as the main source of energy. Conceptions such as the classic contraction theory or that of crustal convection currents have a thermal or thermodynamic basis.

However, the primary stock of internal heat appears to be insufficient to explain the long terrestrial history with repeated periods of orogenesis. The other source of internal heat, viz. decay of radioactive elements, may have been most important in the early turbulent days of the formation of our planet; but the present terrestrial stock of radioactive elements is only an unimportant remnant of our inheritance of natural radioactivity in the stellar matter, from which our planetary system originated.

That there are still natural radioactive elements in our time, more than three thousand millions of years after the origin of our planet, is only due to the very slow rate ot disintegration of some of the intermediate elements of the radioactive series. So it seems that thermal processes (either cooling or radioactive heating) provide insufficient amounts of energy for the motor of endogenic evolution. Diastrophism and volcanism are apparently still as active in our day as in late pre-Cambrian times.

There are still other objections against the thermal convection currents theory, such as its premise of chemical homogeneity of the substratum, which is at variance with modern cosmochemical and geological views, as well as with seismological observations (VAN BEMMELEN, 1952).

Therefore, geophysical theories which take into account only thermal scources of endogenic energy cannot explain the manifold aspects and long duration of geological evolution. They face the same difficulty as that which confronted astrophysics before the

9

development of our knowledge of atomic energy, viz. inadequacy of the known source of energy.

There must be other sources of endogenic energy, beside the primeval internal heat and radioactivity. The earth apparently contains stores of energy of enormous potentiality, which are more or less gradually released during the lifetime of our planet, periodically causing cycles of orogenesis and volcanic activity.

The science of geophysics is at present somewhat at the same level of evolution as astrophysics some decades ago. It was formerly thought that the contraction of stellar gases could provide the energy for the evolution of the stars according to RUSSELL's diagram. However, this source of energy appeared to be utterly inadequate, when the long life of such stars as our sun was taken into consideration. A way out of this difficulty was provided by atomic energy, viz. the liberation of energy by nuclear reactions (transmutation of mass into energy of radiation). So astrophysics evolved from thermodynamic conceptions to theories of nuclear physics.

Owing to the fact that our planetary evolution occurs at a much lower energy level than does stellar evolution, transmutation energy generated by nuclear reactions cannot form this additional source of energy. Instead, our planet is a huge physico-chemical system, the internal structure of which is determined by interatomic forces. Reactions between the electronic shells of the terrestrial elements can be considered as the main source of endogenic energy.

Recently, KOBER (1951, 1952) has evolved the contraction theory by his ,,atom geology'', suggesting that in the core of our planet the formation of trans-uranium elements still proceeds. The present author is of the opinion that the planetary energy level is too low for such nuclear reactions. Instead, the present author holds that the science of geophysics will move on from thermo-dynamical to physico-chemical conceptions (VAN BEMMELEN, 1948).

The earth is a cosmic body which has no internal equilibrium. Our planet contains reserves of free energy which cause geological evolution as a subordinate part of the general evolution of the planetary system to which it belongs. It is hotter than interstellar space, and it may contain in its heavy core solar gases, entrapped during the creation of our planetary system.

Interatomic attractions and repulsions form a powerful store of internal energy. These forces are generally locked in balanced physico-chemical systems. Once thrown out of equilibrium (by a change of the conditions of temperature and pressure) the inter-atomic forces are capable of producing any effect of orogeny, for the ensuing chemical reactions are accompanied by the migration of constituents and important changes of specific density and other physical properties.

These geochemical reactions will disturb the gravitational (hydrostatic) equilibrium in the silicate mantle and, finally, mass displacements will occur, which tend to restore the hydrostatic equilibrium. At the surface such processes will be reflected as differential vertical movements (primary tectogenesis), which are accompanied by temporary deviations from isostasy, besides normal intermediate and deep-focus earthquakes. This primary tectogenesis creates an orogenic relief, containing in its turn accumulations of potential energy, which will give rise to gravitational reactions (called secondary tectogenesis).

This conception is supported by observations in deeply eroded older mountain belts, which, at present form parts of consolidated continental shields. The deeper levels of the crust have been transformed by geo-chemical processes of migmatization and subsequently pushed up and intruded by bodies of granitic magma. This gives rise to the mantled gneiss domes, some examples of which are described by PENTI ESKOLA (1948). The repeated upheaval of such plutonic complexes is an established fact, that throws light upon the nature of orogenesis. It seems that the rising granitic magma has supplied the elevation power. What, then, gives the magma its power to move upward, and to lift its cover? And what is the explanation of the universal concentration of granitic magma in orogenic zones? The answer to the first question is most probably the lesser density of the granitic magma as compared with the average crystalline rocks. The answer to the second question might be that beneath geosynclinal belts geochemical processes are active, which give rise to accumulations of juvenile granitic magma, called neo-sial. These accumulations corrode the overlying crust by the process of migmatization, and — after having attained critical dimensions — they are subsequently pushed up, creating crustal bulges at the surface. Such bulges may occupy the central part of a basin of subsidence; or they may be aligned in geosynclinal belts, forming mountain systems of Alpine character.

It is not necessary that the orogenic uplift is related to only one, huge "magmatic blister", as was suggested by J. L. RICH (1951). The study of mountain systems of Alpine character has revealed that they were formed by a series of orogenic phases, which occurred in parallel belts, shifting step by step sidewards, from the central parts of the mobile belt toward the foreland. So we have to deal with a series of upwarps, causing more or less arcuate mountain belts or rows of islands, such as the Alpine system or the East Asiatic island festoons. This principle of lateral shift of the process of mountain building in the course of the geological evolution will be demonstrated in the second part of this book for the "case history" of the Sunda area in Indonesia.

E. SUESS is the author of the phrase "Entgasung der Erde" (degassing of the earth), thus indicating the basis of his contraction theory. This upward and outward migration of gases was always a fundamental principle of the French school of petrologists (TERMIER, LACROIX, etc.), who spoke of "emanations". KUHN and RITTMANN (1941) even suggested that the entire core of the earth might consist of solar gases in a highly compressed, supercritical state. The present author (1948) followed a more moderate course in agreeing that the heavy metallic core of our planet might contain large amounts of solar gases, e.g. hydrogen. R. A. DALY recently discussed the possibility that the escape of juvenile gases is helping to organize the earth spheres, like an engine working through geological time. This idea was also taken up by TOM F. W. BARTH (1952), who suggested that the degassing of the earth can be considered as the main source of endogenic energy. BARTH is of the opinion that such a degassing would cause a shrinking of the earth, so that his theory would be another variant of the contraction theory. The present author, however, thinks that exothermic physico-chemical chain reactions will ensue, and that the energy, released by such a redistribution of matter in the concentric shells of the earth, is the fundamental source of endogenic energy.

The earth evidently contains enormous reserves of energy which are released more or less gradually. The rise of energy from the depth and its ultimate radiation into space is a long and complicated chain-reaction, during which the energy assumes various forms, e.g. physico-chemical, gravitational, thermal. All processes are subjected to the second main law of thermo-dynamics and so they strive to maintain a state of equilibrium in the interior of the earth.

Chemical reactions in a restricted part will disturb the hydrostatic equilibrium with respect to its surroundings. The ensuing mass displacements change the local gradients of pressure and temperature and will disturb in their turn the chemical equilibrium in the adjacent parts, and so on. In this way, these chain-reactions spread, step by step, from a primary focal area.

There are no closed physico-chemical systems in nature; the chain-reactions will go on pulsating at a varying rate. This ,,pulse of the earth'' is an outward sign of the ageing of our planet. When it stops for lack of internal energy, crustal evolution — and with it, human life on earth — will come to an end.

Such internal geochemical adjustments must cause a progressive arrangement of the terrestrial elements in concentric spheres. The inner layers — like the core at more than 2900 km depth — might be primary features of the earth. They might be formed during the cosmogony. In the outer part of the silicate mantle of the earth called tectosphere, such adjustments were hampered by its high viscosity, and most probably some undercooling occurred at the outset. The retarded release of this geochemical energy can provide a fully adequate source of energy for endogenic geological evolution (VAN BEMMELEN, 1948, 1952).

Relations in theories on mountain building

In current theories regarding mountain building the following combinations of the main principles are generally met with:

a) The problem of forces. Geotectonic theories are either "unicausal-istic", accepting tangential compression of the crust as the main cause of mountain building; or they are "bicausalistic", distinguishing between primary vertical forces and secondary horizontal spreading by gravity.

Of course, the primary differential vertical movements might be the result of various endogenic causes. Differential cooling of the underlying columns might cause greater shrinking of one column with respect to others. Differential heating, for instance by local concentrations of radioactive matter, might cause a swelling up and the formation of magmatic "blisters" (RICH, 1951).

Restoration of the isostatic equilibrium may be another cause of differential vertical movements. For instance, the rise of Scandinavia after the melting of the Pleistocene ice cap; or the uplift of a mountain range after the formation of a mountain root in the sense of AIRY, during a previous phase of crustal compression. Such mountain roots may also originate by other processes, viz. of a chemical nature.

Geological processes connected with long-range migrations of constituents will lead to changes of the composition, which are accompanied by changes of density and bulk of parts of the underlying columns of silicate mixtures.

Such physical and chemical processes in depth will affect the density of the matter concerned. Therefore, due to the differences in the intensity and speed of these processes, the hydrostatic equilibrium will be disturbed in depth. Finally, mass displacements may occur, which tend to restore the hydrostatic equilibrium in depth. At the surface these subcrustal mass displacements will be reflected as differential vertical movements ("primary tectogenesis").

Such differential vertical movements of the crust and its sedimentary cover cause, in their turn, a field of potential energy, which may cause gravitational reactions ("secondary tectogenesis").

In the unicausal group of theories, plutonic and volcanic events are considered independent from or as a byproduct of mountain building by compression. In the bicausal group, magmatic processes are granted a more primary rôle in the creation of the orogenic relief. Other, poly-causalistic conceptions are also possible, accepting endogenic causes of primary tectogenesis, as well as external cosmic disturbances, for instance if passing cosmic bodies cause disturbancies in the earth's crust; but these external influences are not considered here, being less probable.

b) *The problem of movements.* Moreover, geotectonic theories are either "mobilistic", accepting continental drift to a greater or lesser extent, and accepting a narrowing of the mobile, geosynclinal belt, which leads to a concentration of sialic matter in that belt during the process of mountain building; or they are "fixistic", keeping the continental shields more or less rooted and fixed in position in relation to the main mass of the geoid, so that during the orogenesis no concentration of sial from the sides occurs in the mobile belt.

KOBER'S bench-vice concept — i.e. the sqeezing out of the geosynclinal matter between the rigid blocks of fore- and hinterland —

14

leads either to acceptance of the old contraction theory, or to belief in continental displacements caused by the viscous drag of subcrustal convection currents. The same can be said of VENING MEINESZ' theory of crustal buckling, UMBGROVE's idea about underthrusting of the fore- and hinterland blocks as the cause of Alpine orogenesis, AMPFERER's suction theory (Verschluckung) and KRAUS' "Abbau" theory. If it is accepted that the crustal shortening by contraction is insufficient to explain the repeated crustal shortenings in orogenic belts from Precambrian times up to the present, the only solution for the unicausal conceptions of mountain building is lateral displacement of large crustal blocks with respect to the bulk of the geoid. In other words, "unicausalism", if not linked with contraction of our planet, leads to "mobilism" of the sialic crust.

On the other hand, fixistic conceptions of crustal evolution, if not based on the contraction theory, lead to a bicausal interpretation of orogenesis. For in this case the cause of the orogenic deformations is not sought in displacements of fore- and hinterland, but in the· geosynclinal area itself. Mountain building in the geosynclinal area will be associated with movements of magmatic matter underneath, which create the orogenic relief in the geosynclinal belt (primary tectogenesis); whereas folding and overthrusting are considered to be gravitational reactions to this primary relief (secondary tectogenesis). The fore- and hinterland are not influenced by the orogenic process; at least not before its final phases, when waves of crustal deformations (undations, geanticlinal uplifts and their foredeeps), starting from the centres of diastrophism in the mobile belts, come to a halt at the borders of the adjacent continental shields. In the older crustal blocks previous periods of mountain building and accompanying magmatic processes have already more or less exhausted the available endogenic energy.

So it appears that "bicausal" conceptions of the forces in mountain building are generally related to "fixistic" theories of geotectonic movements.

c) The problem of energy. The source of endogenic energy is sought in *thermal* or in *chemical* processes. The main difference between the two possibilities is that the thermal energy system is based on reversible changes of density (cooling and heating), and the chemical energy system on irreversible chain-reactions (changes of composition).

Subcrustal convection currents need reversible expansions and contractions of matter. This is an essential premise, a "conditio sine qua non" for repeated convection currents in the substratum. So they depend on a thermal energy system. Consequently, the uni-causal-mobilistic trend in theories on mountain building logically points to a thermal source of energy (either general cooling of the earth, or local and periodical radioactive heating).

The bicausal-fixistic school, however, tends to consider magmatic processes underneath the mobile (geosynclinal) belts as the main source of endogenic energy. KRAUS (1928) tended to adhere to this school in his earlier publications, but in his recent books on orogenesis he stresses the importance of convection currents, leaving only a secondary role to hypodifferentiation (1951).

In general, however, the bicausal-fixistic school considers geochemical processes as being the fundamental sources of energy. These processes cause irreversible changes of density.

So we come to the conclusion that in current theories of mountain building the main trends of thought might be grouped in *two classes*: *I) the unicausal-mobilistic theories, with a thermal source of energy, and II) the bicausal-fixistic theories, with a chemical source of energy.*

The first class is favoured by the majority of contemporary geologists and geophysicists.

The second class has not yet been widely accepted as a basic conception of orogenesis. It is advocated by the author since 1931. BARTH is inclined to it in his paper of 1952. Recently, UMBGROVE admitted that changes of chemical composition take place in the mantle down to great depth (1952, p. 108). According to a note in the "Bulletin Analitique du C.N.R.S. de France (Vol. 13, 2, 1952 nr 13. 126 on p. 37)", the Russian author BELOUSOV is also of the opinion that geochemical differentiation of the silicate mantle is the fundamental process for the tectonic evolution of the earth. The geochemical concept agrees with modern developments in petrology (migmatization, granitization, transformation, diffusion of matter due to geochemical gradients). This second class of orogenic theories will most probably come to the fore in the near future and its geological and geophysical aspects will then require serious attention.

Basic principles of the undation theory.

The present author classifies his own geological conceptions under the aforementioned second class (bicausal-fixistic with a geochemical source of energy).

About 25 years ago he started work on the Geological Survey of Indonesia as an adherent of the first class, having written a doctoral thesis of the nappe-structures of the Betic system in southern Spain, which was based on unicausal and mobilistic conceptions. But soon afterwards field observations in Indonesia proved to be at variance with deductions from this line of thought. New inductive grouping of the primary stratigraphical, volcanological, tectonic, and geophysical data led him to the conclusion that the bicausal interpretation of the orogenic phenomena and a fixistic conception about the position of the continental framework of Indonesia provided a more satisfactory basis for the description of its geological evolution.

After the second world war the Government of the former Netherlands East Indies commissioned the author to compile a synthesis of the geological knowledge of Indonesia, obtained during the period — about one hundred years — of the existence of the Netherlands Indies Department of Mines and Geological Survey (VAN BEMMELEN 1949).

The manner of presentation and interpretation of the basic facts in this synthesis differs at many essential points from the conception of other authors, such as BROUWER, VENING MEINESZ and UMBGROVE, who based their general pictures on unicausalistic-mobilistic ideas.

RUTTEN's lectures on the geology of the Netherlands Indies, which appeared in the Dutch language in 1927, are of an analytical character, giving the most valuable presentation of our geological knowledge at that time.

In the author's "Geology of Indonesia" full use has been made of RUTTEN's excellent critical analysis of the facts and of later data; but more stress has been laid on the synthesis of the data, in accordance with the instructions of the Dutch Government, who wished to round off the results of one century of research.

In the second part of this book an attempt is made to outline the results of this synthesis and to indicate how it has been attained by alternative inductions and deductions.

As a synthetic picture necessarily implies and applies basic principles, it is desirable to elucidate those premises here. The general

17

framework for the description of the geological evolution of Indonesia is provided by the author's undation theory, which is based on two fundamental conceptions. These are (I) *the geochemical concept of crustal evolution* and (II) *the bicausality concept of mountain building.*

I. *The geochemical concept of crustal evolution.* According to the first concept the source of endogenic energy is to be sought in physico-chemical chain-reactions in depth, which, as a whole, are irreversible in accordance with the second main law of thermodynamics. Such chain-reactions cause the reserves of the earth's physico-chemical potential energy to be used for the distribution and organisation of the primeval planetary matter into the shells of the earth. The crustal growth and evolution is conceived as the results of the splitting up of the intermediate basaltic layer of the tectosphere into a simatic and a sialic fraction. The segregation of sima and sial from the intermediate layer of basalt can be compared with the segregation of the cambium layer of a tree which produces an outer crust (bark) and an inner structure (wood fibres).

It is a well established rule that there is a systematic relation between the stages of orogenic evolution and the accompanying consanguinic suites of igneous rocks (STILLE 1940, VAN BEMMELEN 1950, 1953).

The *Atlantic suite* occurs outside the orogenic belts, or in older parts of the crust which broke down into more mobile geosynclinal belts before the orogenic cycle had started.

The *ophiolite suite* of basic and ultra-basic igneous rocks is characteristic for the geosynclinal stage of subsidence.

The *Pacific suite* is formed during the stage of uplift of mountain ranges or island arcs from the geosynclinal sea, i.e. during the orogenesis in a stricter sense. The plutonic rocks, occupying the core of the geanticlines are acid (granitic) in composition; whereas the effusive volcanic rocks, occurring in higher crustal positions, are on the average more basic (basalto-andesitic). This observation appears to be at variance with the conventional gravitational crystallisation differentiation which considers granitic magmas as the rest melts which accumulate on top of the parental basaltic magmas.

The calc-alkaline Pacific suite of volcanites obtains in later stages

18

of the orogenic evolution the more potassic character of the *Mediter-ranean suite*. There are strong indications that this change of the chemical composition of the Pacific eruptive rocks is due to the assimilation of large quantities of limestone.

After a cycle of orogenesis, during which the crust has been trans-formed and invaded by these successive suites of igneous rocks, it may be finally reconsolidated into a continental shield. In post-orogenic times such reconsolidated crustal parts are sometimes subjected to effusions of *plateau basalts*, which are issued as fissure eruptions from a magma layer poor in gases and situated at the base of the sialic crust.

The time-relations between suites of igneous rocks and phases of orogenic evolution, and observations about the geometric relations between contemporary basic and acid igneous rocks in the orogenic structures, can tentatively be interpreted according to the geoche-mical concept, as follows:

The intermediate basaltic layer of the tectosphere, situated at the base of the granitic crust and above the seismic Mohorovičić discontinuity (about - 40 km), is considered to be the parental magma layer (see fig. 31). Splitting up of this magma by some process of chemical differentiation would produce granitic (sialic) magma on top and a simatic residue at the base. Therefore, this parental magma might be called sialsima (or salsima, or sialma). It can reach the surface in a more or less uncontaminated form only outside the orogenic belts. With RITTMANN the author is of the opinion that the basalts of the sodic Atlantic suite are closely related to the parental salsima.

Study of the evolution of the earth's crust has shown that period-ically, at intervals of some hundreds of millions of years, geosyn-clinal areas of subsidence (mobile belts) develop, which become the birthplace of a mountain system. In Europe, for instance, the Cale-donian, Hercynian and Alpine systems can be distinguished. The subsidence of such geosynclinal seas, like the Tethys sea from which the Alpine system originated, is probably volumetrically more or less compensated by the rise of the adjacent crustal tracts, from which the geosynclinal sediments are derived. Such large-scale differential vertical movements, affecting areas of more than 1000 km in diameter, are called geo-undations by STILLE and the author. They are presumably the result of geochemical chain-reactions which ultimately change the specific density of extensive parts of the substratum to such an extent that deep seated and widespread mass-displacements occurred, which strove after restoration of hydrostatic equilibrium.

The geochemical processes causing the rise and subsidence of extensive crustal tracts (geo-undations) at intervals of some hundreds of millions of years, are of a different nature and presumably more deep seated than the differentiation of the salsimatic basalt magma accompanying the formation of a mountain system. Suggestions about the character of such deep seated processes, causing geo-undations, are of necessity very speculative. The author puts forward such a possibility in § IX of his paper "Cosmogony and Geochemistry" (1948), viz. the regeneration of the ultrabasic sima layer (situated at the depth of between 40 and about 100 km) into basic (basaltic) salsima. Such a "hypo-migmatization" of the sima might be due to the outward migration of lithophile and atmophile constituents from deeper levels of the silicate mantle (emanations) or even from the core of the earth which contains entrapped solar gases (BARTH's principle of degassing of the earth).
Regeneration of the basaltic salsima layer would cause a tumefaction of that part of the tectosphere and the outflow of plateau basalts at the surface. Thereafter, cooling and consolidation would follow, accompanied by the breaking down of extensive crustal tracts, loaded by thousands of metres of plateau basalts. In this way extensive parts of the Atlantic and Indian Oceanic basins may have come into existence. (STILLE's "Neu-Ozeane", 1948).

In this book, however, we shall not discuss these geo-undations, these uplifts and downwarps of continents, nor the formation of mobile belts like the Tethys. We shall restrict our attention to the process of mountain building which takes place inside the limits of geosynclinal belts, such as that resulting in the Alpine and Indonesian mountain systems, which developed from the Tethys belt. This process of mountain building appears to occur in the following way: Within the geosynclinal area, after a long period of quiet subsidence and sedimentation, a median geanticline is pushed up in the central deepest part. This uplift is volumetrically compensated by sidedeeps. From this embryonic stage the mountain system develops further by the pushing up of the sidedeeps into geanticlines, and the displacement of the sidedeep (or foredeep) toward the foreland.

Such geanticlines and their geosynclinal foredeeps are wave-like deformations of the earth's crust with a width of 100–200 km, called undations in a stricter sense (Spezialundationen" by STILLE, "meso-undations" by the author). They are formed at intervals of some dozens of millions of years. Thus the process of orogenesis appears to spread from the central and deepest parts of the geosynclinal mobile belt (the foci of diastrophism) towards the foreland, where it finally comes to a standstill. Historic-geological analyses of various mountain systems originating from mobile belts show this general scheme. It has, for instance, been demonstrated by STILLE in the case of the Hercynian system in Europe, by the author (1933) in that of the Alpine system in Europe, and in this book examples from the Indonesian archipelago are given (Sunda area, Banda arcs).

The formation of mountain systems in regions which were previously subjected to geosynclinal subsidence can be explained according to the undation theory in the following way.

During geosynclinal subsidence the chemical balance of the intermediate basalt layer of the tectosphere (the "salsima") is disturbed. Gradients of pressure and temperature are changed, and in those places where these changes were most important (i.e. under the deepest and central parts of the geosynclinal sea) geochemical migrations of certain elements are initiated. The rise of lithophile and atmophile constituents will cause segregation or "hypo-differentiation" of the basaltic layer, by which matter of lower density accumulates in its upper part, whilst the residue in the lower part will increase in density. Thus, such a slow process of hypo-differentiation, which is a reaction to the geosynclinal subsidence, will split up the salsima layer with a specific gravity of approx. 3 into accumulations of sialic magma of a specific gravity of less than 2.7 and simatic matter of a specific gravity of 3.3 or more. So, after a certain incubation period, the geochemical migrations (which tended to restore the physico-chemical equilibrium in a restricted part of the salsima), disturbed the hydrostatic equilibrium between that part and the surrounding salsima. The total volume of the part of the salsima layer subjected to this process of hypo-differentiation remained more or less the same. There may be some expansions due to the positive heat balance of the chemical reactions; for the latter will be exothermic, being a reaction to cooling. Apart from changes of temperature, changes of chemical bonds will also be accompanied

21

by some changes of density and volume. But on the whole, the hypo-differentiation of the intermediate basalt layer is a slow and quiet process, going on in the depth without giving rise to important differential vertical movements at the surface. In the long run, the intermediate salsima layer is segregated into an upper root of neo-sial (called "asthenolith") and an antiroot of neo-sima below (see fig. 10 of part II).

The sima beneath the Mohorovičić discontinuity may have a peridotitic composition, with a specific gravity of about 3.3. Neither this sima, nor the neo-sima accumulating on top of it, have any relations whatsoever with the ultrabasic peridotites of the ophiolite suite, which are emplaced at the bottom of the geosynclinal sea. The latter are formed by a quite different process, viz. the basification and transformation of crustal rocks and geosynclinal sediments by Mg, Ca and Fe ions, chased out, and swept together by the ascending migmatite front.

The ophiolites can be considered as the products of a basic front in the geosynclinal stage. The juvenile granitic magma, accumulating at the base of the crystalline crust during the geosynclinal period, corrodes and migmatizes the overlying crust by its emanations. This migmatization of the lower part of the crust causes its acidification into a migma of granitic composition. The calcium-, iron-, and magnesium-constituents, which do not fit into the chemical bonds of this zone, are chased outward and upward. The hydrothermal solutions of these cafemic constituents enter into the higher crustal levels, where they may react with the country rock, or, if they do not encounter mineral aggregates in which they are fixed by chemical reactions, they may reach the floor of the geosynclinal ocean as submarine thermal mineral springs. Such rising solutions of the basic constituents of the lower part of the crust may cause phenomena such as dolomitization, chloritization, serpentinization, amphibolitization, etcetera. The transformations of the country rock may lead to ultra-metamorphism, forming basic and ultrabasic igneous rocks and palingenic magmas of the ophiolite suite. Such a geosynclinal basic front is complimentary to the acidification of the lower part of the crust by migmatization; however, unlike the migma-zone, it is not always a coherent zone, because it is more dependent on the chemistry of the country rocks. It may appear as large masses of peridotite and serpentinite, but it may also appear as scattered lenses of amphibolites and the like. In general, it need not border directly upon the acid migmatite front.

The ophiolites, formed at the base of the geosynclinal sea and reaching its floor, are contemporaneous with the neo-sima, formed at the base of the intermediate salsima layer. However, genetically, they have nothing in common. The ophiolites are separated from the ultrabasic neo-sima by a granitic asthenolith and what remains of the intermediate layer of parental basalt after its hypo-differentiation. It is geochemically and mechanically quite improbable that the ophiolites, intruded and extruded during the geosynclinal stage, are derived from the ultrabasic rocks at the base of the tectosphere, below the Mohorovičić discontinuity.

II. *The bicausality concept of mountain building.* Let us now consider the effect of the disturbance of hydrostatic equilibrium in the salsima layer by its splitting up into a neo-sialic root and a neo-simatic antiroot. The latter will have a tendency to sink down in the simatic layer and then spread out. Owing to this process, the Mohorovičić discontinuity will first be pushed down, and will then straighten out again parallel to the surface of the geoid. The restoration of hydrostatic equilibrium in these deeper parts of the tectosphere is hydrodynamically independent of the mass displacements tending to restore the hydrostatic equilibrium in the higher parts. Consequently, the deformations of the basic antiroot and the acid mountain root will, most probably occur at different rates. This will effect the gravity field at the surface; temporary deviations from the isostatic equilibrium will be observable. The negative isostatic anomalies above nascent mountain ranges can be explained in this way.

The root of neo-sial (juvenile granitic matter) grows upward by migmatic accretions, which corrode the overlying sialic crust. This body of acid magma and migma, called "asthenolith" by the author, will have a relatively high temperature due to the positive heat balance of the exothermic physico-chemical chain-reactions by which it was formed. Consequently, its viscosity will be lower and its mobility higher than that of the overlying crust. Moreover, in its active stage of development, this asthenolith will be saturated with juvenile and resurgent volatile constituents.

An asthenolith is a relatively hot and gas-laden body of magma and migma with low viscosity and low specific gravity (2.4–2.6), which is pushed upward by Archimedean forces. Only after its degassing and consolidation to granitic and migmatic rocks an asthenolith will assume a rigidity and specific gravity comparable to that of the older parts of the sialic crust (2.7).

In the first stages of its formation the asthenolith is too light with respect to its surroundings and tends to float upward. When certain limits of strain are exceeded it will be pressed upward. The floor of the geosyncline is arched up by this rise of the asthenolith and a median geanticline or domal rise is formed.

At this stage offshoots from the asthenolith may diapirically invade the overlying strata, or constituents driven from the acid migmatite front may transform the overlying rocks into high-level, orogenic basic fronts. In the latter case, palingenic magma types of the calc-alkaline Pacific suite of igneous rocks are generated, which are the source of the orogenic volcanism at the top of the geanticlinal arches.

According to this geochemical concept, the roots of mountains are not the result of the pushing down or sucking down of the pre-existing sialic crust, as is held by the unicausal-mobilistic theories. Instead, the idea is suggested that these mountain roots accumulated from below by natural geochemical processes which are a reaction to geosynclinal subsidence. They are composed of juvenile magmatic matter as well as migmatic matter.

The asthenolithic roots are ultimately pushed up by hydrodynamic processes which strive after a restoration of hydrostatic equilibrium.

This uplift of an asthenolith causes the arching up of a geanticline, whereas the adjacent belts will subside on account of volumetric compensation forming side- or foredeeps. Salsima from the foredeep zone accrues towards the zone of the rising asthenolith, thus closing the hydrodynamic circuit. The thinning of the salsima layer beneath the foredeep changes the gradients of temperature and pressure in that belt to such an extent that the process of hypo-differentiation is also initiated under the foredeep.

After an incubation period of some dozens of millions of years this foredeep zone, in its turn, is pushed up by an asthenolith. Thus a new geanticline is formed, whilst the axis of the foredeep shifts another step farther outward, migrating towards the foreland. In this way mountain systems develop from a geosynclinal sea, by way of systems of crustal waves which start from foci of diastrophism in the central and deepest parts of the mobile belt and which migrate step by step towards the foreland. These steps represent the successive orogenic phases of a mountain system.

Such undatory deformations of the crustal relief will cause gravitational reactions. They give rise to folding and overthrusting of rock strata in the foredeeps. By the force of gravitation nappe structures can slide forwards towards the foreland in successive stages of orogenesis and the accompanying migration of the foredeeps.

Finally, the piles of nappes are overtaken by the forward shifting orogenic uplift and are exposed in lofty mountain ranges, as in the Alpides of Europe and in the island arcs of the Sunda System of Indonesia. Thus the structure of mountain ranges, which can actually be studied by geologists, can be explained by the second basic principle of the undation theory, viz. the bicausality concept.

According to this bicausality concept the uplift of the mountain ranges and island arcs is considered to be, not the effect of tangential compressive forces, but the result of geochemical processes underneath. In the second part of this book, in the chapter on volcanism, the relations between the volcanic and the orogenic relief are outlined according to this point of view.

We return, to some extent, to the old conception of the plutonists of the 19th century, who held that the underlying magma is responsible for the uplift of mountain ranges. Therefore, in the description of the regional geology of Indonesia, the chapter on volcanism is not put after, but before, the chapter on the tectonic structures. For, according to this basic principle of the undation theory, volcanism is not an accidental by-product of diastrophism. On the contrary, orogenesis is the result of geochemical processes in depth; in other words, orogenesis results from volcanism in its larger sense.

Both, volcanism and diastrophism, are caused by endogenic forces; both are related to the flow of the endogenic energy to the surface. This energy is partly used for an increased entropy of our planet (inner organization of matter in concentric spheres) and the remainder of it is dispersed into the Universe as radiated heat. We may consider the *crustal transformations and accretions* or volcanism in its larger sense (the geochemical migrations of matter from depth towards the surface) as the chemical aspect of this energy flow;

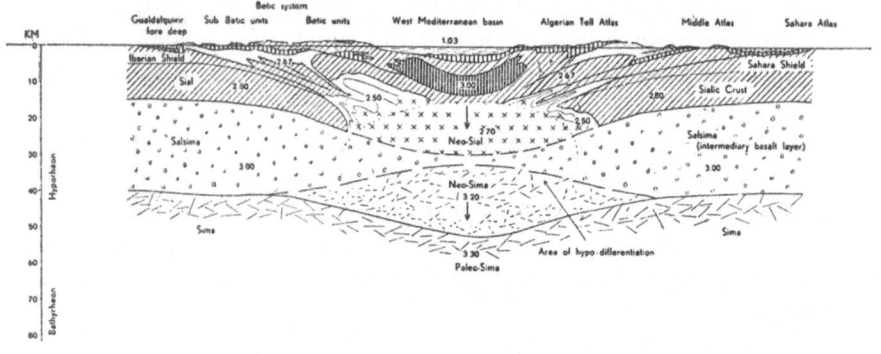

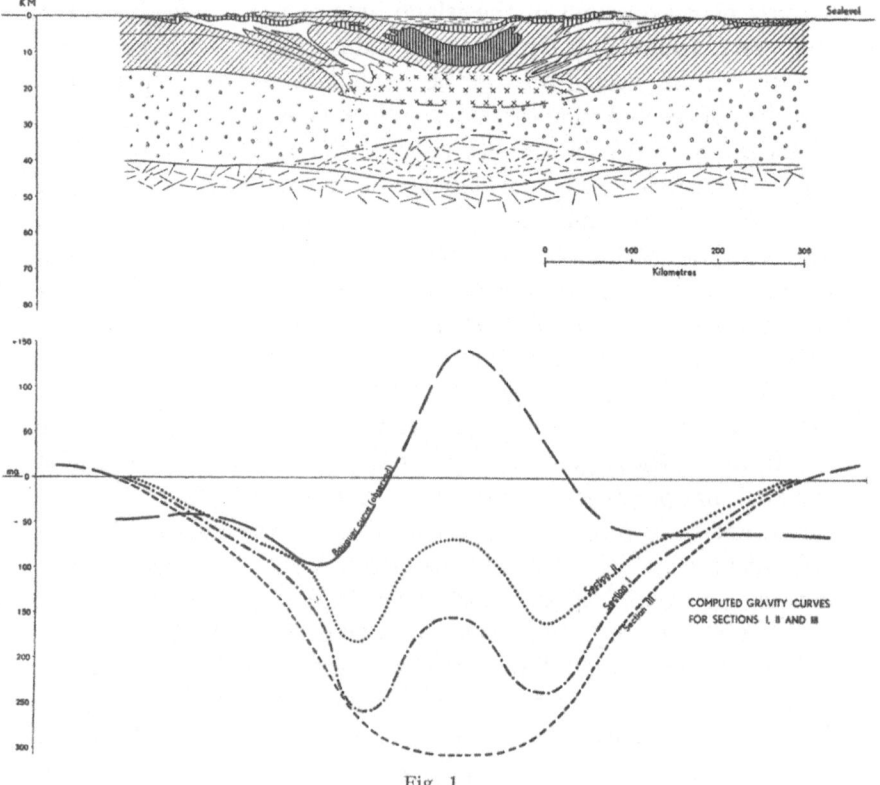

Fig. 1

whilst the *crustal deformations* or tectonics in general represent the mechanical aspect.

Testing the undation theory

In the foregoing pages we have given a picture of the formation of mountain systems according to the basic concepts of the undation theory. The complexity of the process, and the many conjectures involved, leave much doubt about the correctness of these deductions. The reader, probably will wonder whether this theory represents an adequate picture of orogenic evolution. Is it a useful working hypothesis; in other words, is it true?

The value of a theory in Natural Sciences depends on how far it can predict other facts. ARGAND has said: "La valeur d'une théorie est toute entière dans la conformité entre les conséquences qui s'en déduisent et les faits observés".

The test for the orogenic and petrogenic conceptions of the undation theory, outlined in the foregoing pages, lies in the conformity between its deductions regarding observable facts (i.e. its theoretical prognoses) on the one side, and independently determined facts (diagnoses) on the other side. In oil geology, the prognosis of the geologist is tested by the diagnosis of drilling. In the present case, the geological theory deals with the distribution of matter down to depths of dozens of kilometres. In such a case we cannot test a theoretical prognosis by direct observations. Geophysical methods have to be used instead.

Fig. 1

The computed gravity effect of geological sections based on E. KRAUS (*1951*), *according to* HOFMAN (*1952*).

Curve I shows the computed gravity effect of the upper section. The result is a minimum with a high in the centre caused by the ophiolite plate of higher density near the surface.

Curve II shows the computed gravity effect of the second section, which is essentially the same as the first one, only with less geosynclinal subsidence, reducing the gravity minimum.

Due to the moving down of light sialic matter in the central part of the mountain system ("Verschluckung","Abbau"or sucking down of the sialic crust by convection currents), such a theoretical cross section of the Tello-Betic System can never produce a gravity field fitting the observed curve, unless the size of the ophiolite plate is increased to improbable dimensions.

Curve III shows the computed gravity effect of the section across the Tello-Betic System according to the "Verschluckung" or "Abbau" theory, when no hypo-differentiation is taken into account. This curve clearly shows the discrepancy between the prognosis of the gravity field according to this example of the unicausal-mobilistic concept on the one hand, and the diagnostic determination of the gravity on the other.

Geotectonic theories actually suggest structures of the tectosphere and distributions of matter causing definite gravity fields at the surface. Such theoretical predictions of the gravity field can be compared with the observed gravity field, which is independently determined by gravimetric methods.

This theoretical possibility of testing geotectonic conceptions has recently been implemented for the West Mediterranean region.

Throughout a section from the Atlas Meseta in northern Africa to the Spanish Meseta in southern Europe, the geological distribution of the masses was plotted according to the "Abbautheory" of KRAUS and the author's undation theory. B. J. HOFMAN calculated the gravity field which would result from such mass distributions as suggested by those geological sections.

It appeared that the section constructed according to unicausal-mobilistic principles revealed a systematical discrepancy between the geological prognosis and the gravimetrical diagnosis. Buckling of the crust, or its sucking down, means that sialic matter would have been transported to the geosynclinal belt and accumulated in the orogenic system originating from it. Thus, accumulation of sialic matter of low density would cause a regional field of negative isostatic anomalies.

The bench-vice concept of orogenesis, which considers the formation of the Tello-Betic mountain system to be the effect of the sqeezing out of the geosynclinal content between the adjacent crustal blocks of the Iberian and Atlas Meseta appears to have a systematic defect. It leads to the prognosis of a regional field of negative anomalies, due to the presence of a mountain root of low density, whereas gravimetrical diagnosis shows that the West-Mediterranian has a regional field of positive isostatic anomalies.

A mountain root of low density, in the sense of AIRY, formed by a general narrowing of the geosynclinal belt during the process of mountain building, can be removed by subcrustal spreading or by isostatic uplift and denudation at the surface. In the first case the gravimetric effect of the root is not really removed. It is only spread over a wider area, and after the phase of crustal compression the isostatic uplift will occur not only in the compressed zone, but also in the adjacent belts of fore- and hinterland. This prognosis is not confirmed by the study of alpine-type mountain chains, which are always characterized by the presence of sidedeeps. The other solution leads to absurd values of the amounts of matter, which should have been removed by denudation, as was shown by the author for the Swiss Alps (1953).

Therefore, the presence of a mountain root formed by compression of the geosynclinal mobile belt is not confirmed by this gravimetrical test. In a recent paper MINTROP (1953) discussed the problem of mountain roots from a seismic point of view, and this author comes also to the conclusion that neither the Sierra Nevada nor the Alps have mountain roots in the sence of AIRY. Probably, the distribution of masses under mountain belts is more complicated than it is conceived by the unicausal-mobilistic approach of the problem (sucking down, squeezing down, buckling down of crustal matter in a substratum of higher density).

On the other hand, the section across the Tello-Betic system, constructed by the author according to bicausal-fixistic principles, did not show this systematic discrepancy; since no sialic matter is concentrated in the orogenic area from fore- or hinterland, no field of negative isostatic anomalies results. The orogenesis is explained by the splitting up of the intermediate basalt layer into roots of neo-sial and antiroots of neo-sima. The isostatic anomalies are considered to be the result of the incomplete restoration of the hydrostatic equilibrium, which had been disturbed by this hypo-differentiation. The geological section according to the undation theory brings out a close relation between the orogenic relief and the gravity field, such as positive anomalies in the subsiding West-Mediterranian trough and negative anomalies in the rising outer arcs of the Tello-Betic system.

The gravity field predicted by the undation theory appeared to conform perfectly with the gravity field actually observed.

Of course the mass distribution suggested by the undation theory is one of the many possible mass distributions conforming with the observed gravimetric field. But, because of the diametrically opposed methods approach of the gravity field — on the one hand the indirect evaluation by the undation theory which involve a highly complex set of geological inductions and deductions, and on the other the direct measurements of the field by gravimetric methods — such a congruence in the results may be considered to be a test of the correctness of the theory applied. For it means that the theory concerned is capable of correctly predicting the gravity field.

29

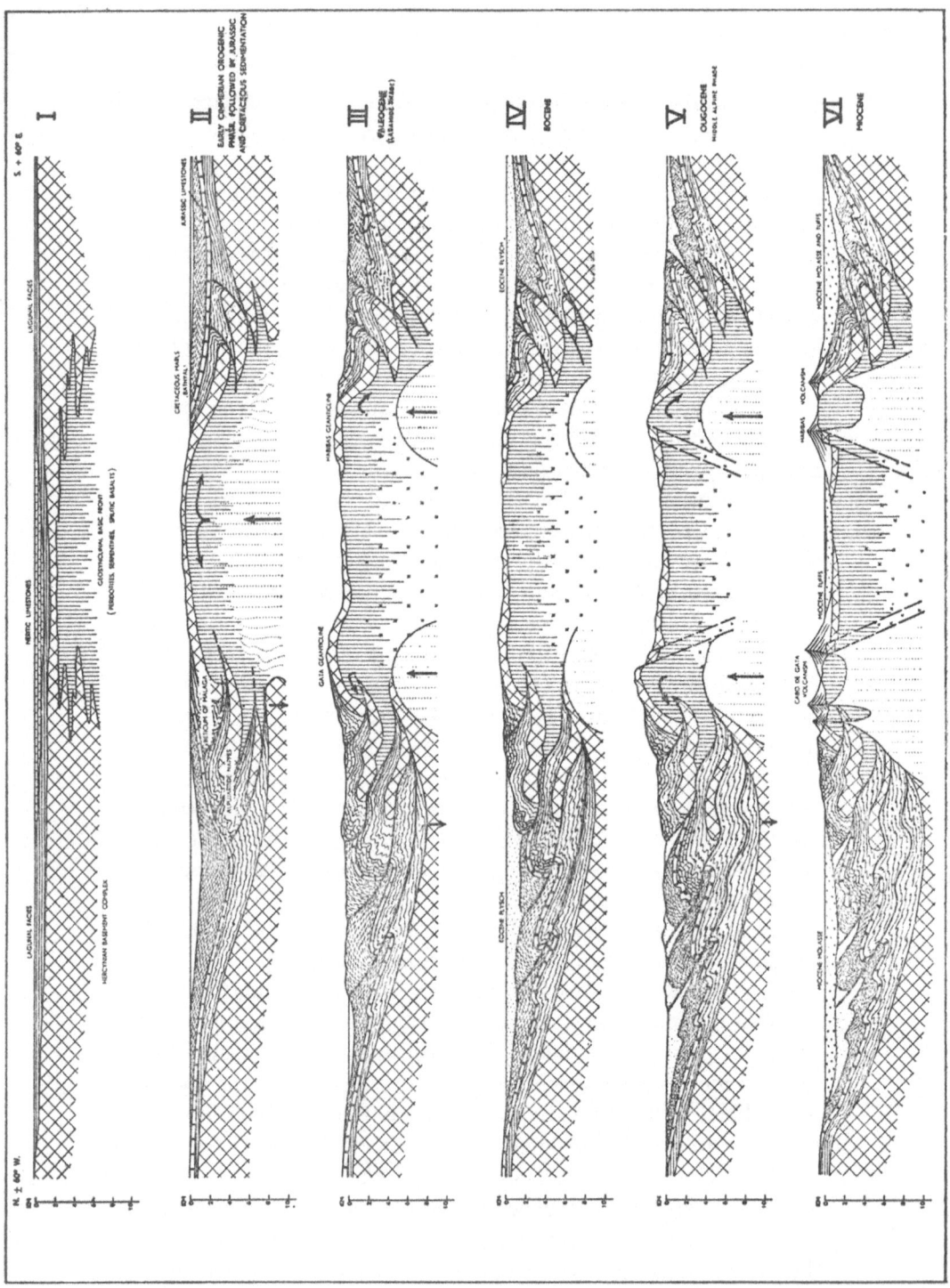

Fig. 2a. The evolution of the Tello-Betic system

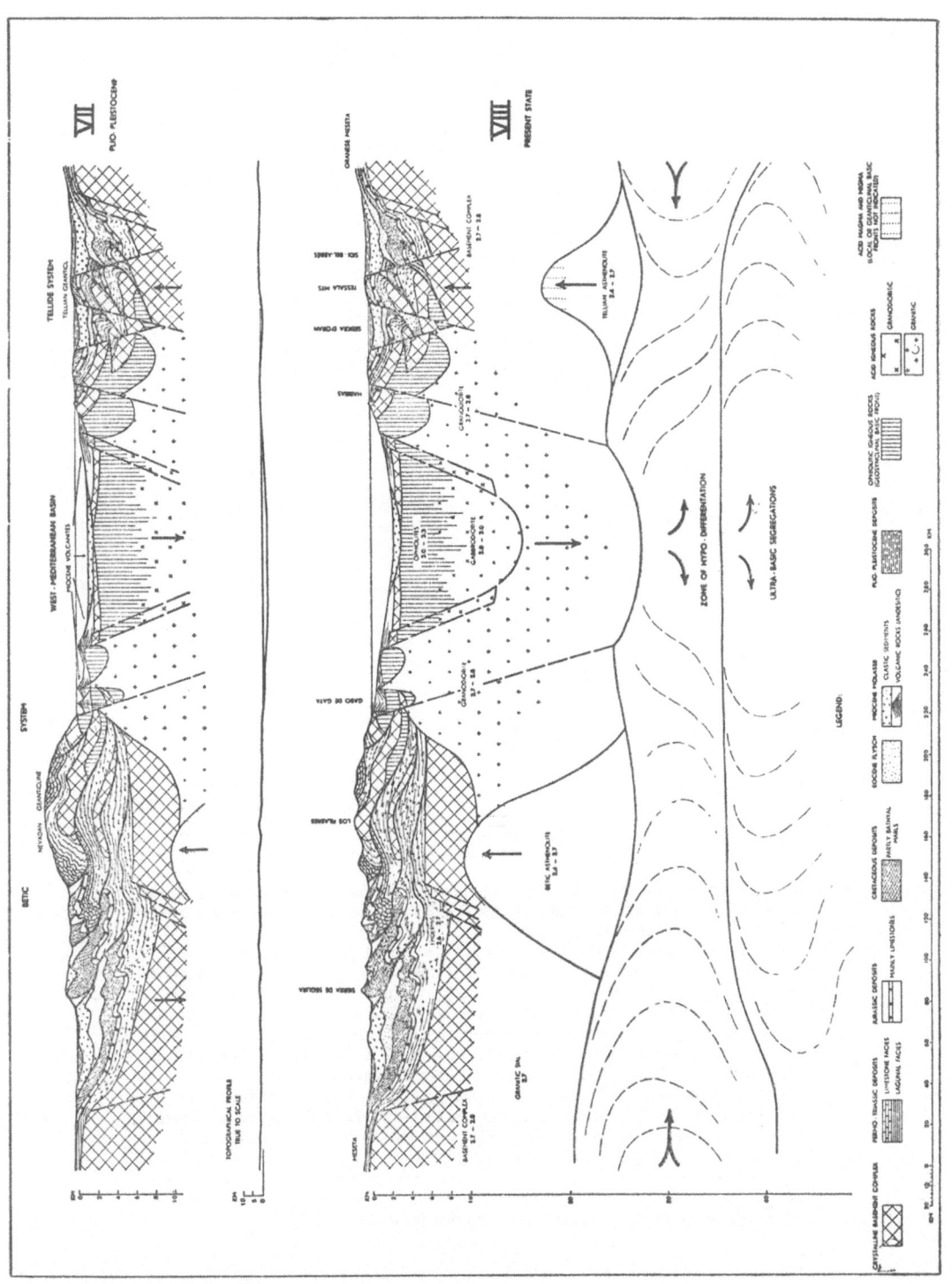

according to the undation theory (VAN BEMMELEN, 1952).

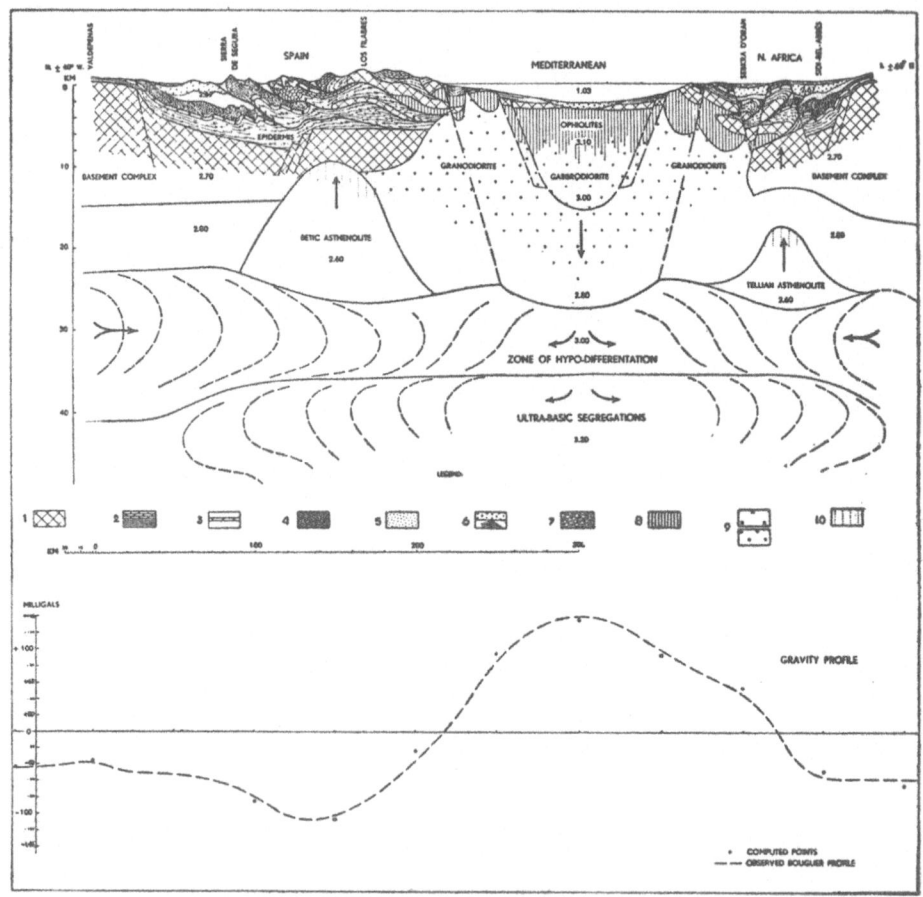

Fig. 2b

Geological section across the Tello-Betic system in the western part of the Mediterranean, according to R. W. van Bemmelen (1952), and its computed gravity effect according to B. J. Hofman (1952).

This geological section is the last of the genetic series of eight sections, showing the evolution of this mountain system from the Triassic up to the present, and constructed according to the bicausalistic-fixistic principles of the undation theory (fig. 2a) The gravity effect of this section was computed by Hofman from specific gravity figures within the limits set a priori by van Bemmelen (see section VIII of fig. 1 in his paper 1952). There appears to be a perfect concordance between this geological prognosis of the gravity field and its geophysical diagnosis..

Legend (see also fig. 2a): 1. *Crystalline basement complex* – 2. *Permo-triassic deposits*, Above: Limestone facies; Below: Lagunal facies – 3. *Jurassic deposits*, mainly limestones – 4. *Cretaceous deposits*, partly bathyal marls – 5. *Eocene flysch* – 6. *Miocene molasse*, Above: Clastic sediments; Below: Volcanic rocks (andesitic) – 7. *Plio-pleistocene deposits* – 8. *Ophiolitic igneous rocks*, (Geosynclinal basic front) – 9. *Acid igneous rocks*, Above: Granodioritic; Below: Granitic – 10. *Acid magma and migma*, (local or geanticlinal basic fronts not indicated)

32

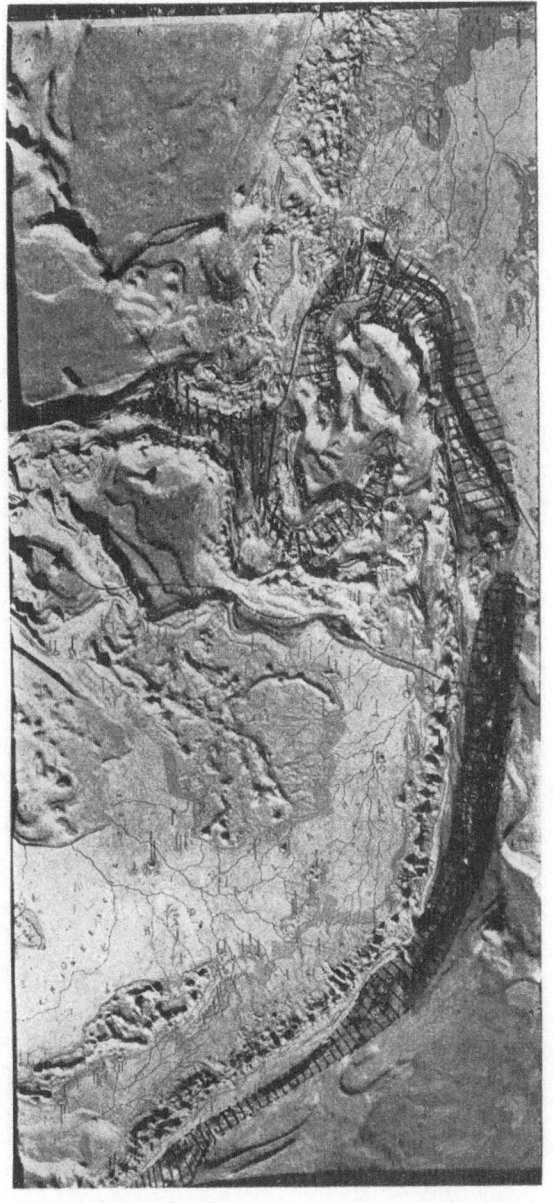

Fig. 3   Relief map of the Indian Archipelago. (The dark coloured belt with transverse shading is the zone of negative isostatic anomalies. The heavy lines are faunistic boundaries, viz. *Wallace's* line and *Weber's* line.

Fig. 4   Photo of the Carstensz group with the South wall cliff glacier. In the back ground the North wall firn. From *Dozy* (1938, table IV).

LITERATURE REFERENCES TO PART I

AMPFERER, O.: Über das Bewegungsbild von Faltengebirgen. Jahrbuch K. K. Geol. Reichsanstalt. Bd. 56, 1906, pp. 539–618.

ARGAND, E.: La tectonique de l'Asie. Comptes Rendus XIII. Sess. Congr. géol. intern. Liège 1922 pp. 171–372 (l.c. p. 293).

BARTH, T. F. W.: Orogeny and Geochemistry. Schweiz. Min. Petr. Mitt. Vol. 32, 2, 1952, pp. 354-360.

BELOUSOV, V. V.: Problèmes de la structure interne de la terre et son évolution. Izvest. Akad. Nank. S.S.S.R. Ser. géogr. 1951, 2, pp. 4–16.

BEMMELEN, R. W. VAN: The bicausality of diastrophism (Undation and gliding). Natuurk. Tijdschr. v. Nederl. Ind. (Journal of Natural Sciences in the Neth. E. Ind.) Vol. 91, nr. 3, Batavia (Djakarta) 1931, pp. 363-413.

———: Über die möglichen Ursachen der Undationen der Erdkruste. Proc. Kon. Ak. v. Wetensch. Amsterdam Vol. 35, nr. 3 1932, pp. 392–399.

———: The undation theory. Natuurk. Tijdschr. Nederl. Ind. Vol. 92, Batavia 1932, pp. 85–242 and 373–403.

———: The undation theory on the development of the earth's crust. Report XVI Int. Geol. Congress, Washington (D.C.) 1933, Vol. 2, pp. 965–982.

———: Die Anwendung der Undationstheorie auf das alpine System in Europa. Proc. Kon. Akad. v. Wetenschap. Vol. 36, 6, Amsterdam, 1933.

———: Das Permanenzproblem nach der Undationstheorie. Geol. Rundschau, 1939, pp. 10–20.

———: Cosmogony and Geochemistry. Proc. XVIII Int. Geol. Congr. Great Britain. 1949, Part. II, London 1948, pp. 9–21.

———: The Geology of Indonesia.
Vol. I.A. General Geology.
Vol. I. B. Plates, literature references, general index.
Vol. II. Economic Geology.
Neth. State Printing Office. 1949. Sole agent: Martinus Nijhoff, The Hague.

———: On the origine of igneous rocks in Indonesia. Geologie en Mijnbouw (J. of the Roy. Neth. Geol. & Mining Soc.) Vol. 12 nr. 7, the Hague, 1950, pp. 207–220.

———: Gravitational Tectogenesis in Indonesia. Geol. en Mijnb. (J. of the Roy. Neth. Geol. & Mining Soc.). Vol. 12, nr. 12, the Hague 1950, pp. 351–361.

———: The endogenic energy of the earth. Amer. J. of Science, Vol. 250, 1952, pp. 104–117.

———: Gravity field and orogenesis in the West-Mediterranean region. Geologie en Mijnbouw (J. of the Roy. Neth. Geol. & Mining Soc.) Vol. 14, nr. 8, the Hague, Aug. 1952, pp. 297–306.

———: Relations entre le volcanisme et la tectogénèse en Indonésie. Bull. Volc., Ser. II, t. XIII, Napoli, 1953, pp. 57–62.

———: Gedanken zur alpinen Gebirgsbildung. Erdölzeitung, Vol. 69, Vienna, Heft 6, 1953, pp. 75–77.

BERTRAND, M.: La chaîne des Alpes et la formation du continent Européen. Bull. Soc. géol. France, 1897.

BORN, A.: Der Aufbau der Erde. Handbuch der Geophysik, Bd. II, 3, Berlin, 1932.

BUCHER, W. H.: Megatectonics and geophysics. Trans. Am. Geoph. Union. Vol. 31, 4, Aug. 1952, pp. 495–507.

CLOOS, H.: Einführung in die Geologie. Berlin, 1936.

ESKOLA, P.: The problem of mantled domes. Qu. J. Geol. Soc. London, Vol. 104, 1948.

GUTENBERG, B.: Hypothesis on the development of the earth. J. of Washing. Ac. of Science. Vol. 20, nr. 2, 1930.

HAARMANN, E.: Die Oszillationstheorie, Stuttgart, 1930.

HARRISON, J. V. and FALCON, N. L.: Collapse structures. Geol. Mag. Vol. 71, nr. 846, 1934, pp. 529-539.

HOFMAN, B. J.: The gravity field of the West-Mediterranean area. Geologie en Mijnbouw. Vol. 14, nr. 8, the Hague, Aug. 1952, pp. 297–306.

HOLMES, A.: Radioaktivität und Geologie. Verh. Naturf. Ges. Vol. 41 Basel, 1930, pp. 136–185.

HUBBERT, KING,: Theory of scale models as applied to the study of geologic structures. Bull. Geol. Soc. Am., Vol. 48, Oct. 1937, pp. 1459–1520.

JOLY, J.: Radioactivity and the surface history of the world. 1924.

KOBER, L.: Atomgeologie. Montan Zeitung, nr. 5, Vienna, 1951, pp. 115–118.

——: Atombau und Geologie. Verh. Geol. Bundesanstalt Vienna, 1952, Sonderheft
C, 7 p.

KRAUS, E.: Das Wachstum der Kontinente nach der Zyklustheorie. Geol. Rundschau
Vol. 19, 5/6, pp. 353–386 & 481–492, 1928.

——: Vergleichende Baugeschichte der Gebirge. Akademie Verlag, Berlin 1951,
Vol. I & II.

KUHN, W. and RITTMANN, A.: Ueber den Zustand des Erdinnern und seine Entstehung aus einem homogenen Urzustand. Geol. Rundschau, Vol. 32, 1941, pp.
215–256.

LANDES, K. K.: Our shrinking Globe. Bull. Geol. Soc. of Am. Vol. 63, 1952, pp. 225–
239.

LEES, G. M.: Foreland folding. Quart. J. Geol. Soc. London, Vol. CVIII, Part. I.
Nr. 429, Dec. 1952.

MINTROP, L.: Die Problematik der Gebirgswurzeln, eine kritische Betrachtung.
Geologische Rundschau, Vol. 41 (Cloos Memorial Vol.), 1953, pp. 67-78.

RICH, J.L.: Origin of compressional mountains and associated phenomena. Bull.
Geol. Soc. Am. Vol. 62, Oct. 1951, pp. 1179–1222.

RITTMANN, A.: Orogénèse et volcanisme. Archives des sciences physiques et naturel-
les, Genève, Vol. 4, fasc. 5, 1951, pp. 273–314.

SCHWINNER, R.: Vulkanismus und Gebirgsbildung. Zeitschr. f. Vulk. Vol. 5, nr. 4,
Berlin 1920.

STAUB, R.: Über die Beziehungen zwischen Alpen und Apennin und die Gestaltung
der Alpinen Leitlinien Europas. Ecl. Geol. Helv. Vol. 44, nr. 1, pp. 29–130, 1951.

STILLE, H.: Grundfragen der vergleichenden Tektonik. Berlin 1924.

——: The growth and decay of continents. Research and Progress, from a lecture
in the Prussian Acad. of Science. Vol. I, nr. 1, Berlin, 1935 p. 9–14.

——: Zur Frage der Herkunft der Magmen. Abh. preuss. Akad. Wiss. 1939, math.
naturw. Kl. 19, Berlin 1940.

——: Ur-und Neuozeane. Abh. deutsch. Akad. d. Wiss. 1945/46, nr. 6, Berlin 1948.

——: Das Leitmotiv der geotektonischen Erdentwicklung. Deutsch. Akad. d. Wiss.
Vorträge und Schriften. Heft 32, Berlin 1949.

——: Das mitteleuropäische variszische Grundgebirge im Bild des Gesamteuropäi-
schen. Beiheft zum Geol. Jahrb. Heft. 2, Herausgeg. v.d. Geol. Landesanst. der
Bundesrep. Deutschland. Hannover, 1951, pp. 138, 2 tables.

SUESS, ED.: Das Antlitz der Erde. Vienna and Leipzig, 1886.

TAYLOR, F. B.: Bearing of the tertiary mountain belts on the origin of the earth's
plan. Bull. Geol. Soc. of Am. Vol. 21 1910, pp. 179–226.

UMBGROVE, J. H. F.: Contraction of the Earth. Proc. Kon. Ned. Ak. v. Wetensch.
Amsterdam, Ser. B, 55, 2, 1952, pp. 105–109.

VENING MEINESZ, F. A.: Maritime gravity survey in the Netherlands East Indies,
tentative interpretation of provisional results. Proc. Kon. Akad. v. Wetensch.
Vol. 33, Amsterdam 1930.

WATERSCHOOT VAN DER GRACHT, W. VAN: Symposium on continental drift. Publ. by
Am. Ass. Petr. Geol., Tulsa, Okl. U.S.A. Thomas Murby, London 1928.

WEGENER, A.: Die Entstehung der Kontinente und Ozeane. 1915.

PART II

THE OROGENIC EVOLUTION OF THE EARTH'S CRUST IN INDONESIA

I. INTRODUCTION

In the last hundred years, and particularly in the years between the two world wars, an impressive body of geological data has been gathered by geologists and mining engineers working in Indonesia. Their total number exceeded one thousand, mainly Netherlanders, but also other Europeans and Americans. They were employed by the East Indian government (the Department of Mining) and private oil and mining companies; moreover some of them carried out their scientific investigations during scientific expeditions and private study tours.

Through these geological investigations, which were often conducted under difficult circumstances in the tropical forests, much has become known about the geological evolution of Indonesia. The resulting ideas on the geological development of this vast region, which occupies approximately 4% of the total area of the earth, has a more than local significance. Here we face the fundamental problems of the origin and evolution of our planet. A thorough understanding of the geological evolution has not only a theoretical, but also a great practical significance, as the geological conditions form the frame within which the history of mankind is enacted.

It may be said that such a general understanding could more easily be obtained from geological observations in better investigated areas, such as Europe and Northern America. It goes without saying that the classical regions of geology have taught us much which applies to Indonesia. But Indonesia itself is a most suitable region for the study of mountain building, because here the orogenesis is still in full swing with all the accompanying phenomena of earth movements, volcanism, anomalies of gravity, and earthquakes. The general relations between these phenomena can be better studied in Indonesia than in more or less completed mountain systems as the European Alps. Because of this active orogenesis, the East Indian Archipelago is one of the most actively volcanic and seismic regions in the world and the relief shows deeps of about 10,830 metres (Mindanao deep, east of the Philippines) and mountains of more than 5,000 metres altitude which rise above the limit of perennial snow in the tropics (the Snow Mountain Range in New Guinea). Nowhere in the world earthquake foci have been found as deep as those under the Flores Sea (at 720 kilometres depth), and gravity anomalies as large as the negative anomaly of 240 milligals (calculated according to the method of regional isostatic reduction) in the Moluccas, are rare elsewhere. (See Fig. 3 and 4).

Geological observations in Indonesia are still widely spaced and the general ideas about the geological structure are consequently imperfect. Yet, some general trends have been established with a fair degree of certainty, and it are exactly these general principles of geological structure and evolution which will lead to a better understanding of the process of mountain building.

The following chapters will deal with the major features of the geological evolution of Indonesia. Also, we shall trace the laws which govern this evolution and consider the conclusions regarding the nature of the evolution of the earth's crust to which this will lead us. This book is therefore not a geological guide, giving detailed geological information on certain areas in Indonesia. The purpose of this book is to stimulate the interest in the geology of Indonesia, both of those who have visited this country or have worked in it, and of those who are interested in the geological history of this island empire from a general scientific point of view.

History of geological research in Indonesia
In 1849 the first state-owned coal mine, (the "Oranje Nassau" mine near Pengaron in Borneo) was opened, and in 1852 the Department of Mining officially recognized as an independant institution. Therefore one century has elapsed since systematic geological research, sponsored by the state, started in Indonesia.

Before that time some research had been done, but geology, as an independant discipline, was still in its infancy. It must be recorded, however, that already at the time of the East Indian Trading Company ("Verenigde Oostindische Compagnie") the famous naturalist G. E. RUMPHIUS carried out important geological investigations in the Moluccas. RUMPHIUS died at Ambon in 1702 and after his death his notes were published at Amsterdam in 1705 under the title of "d'Amboinsche Rariteitenkamer" (the Ambonese Museum of Curiosities).

After the period of British rule in Napoleonic time, the Dutch organized scientific research in Indonesia by founding a Scientific Commission (the "Natuurkundige Commissie"); with this body many enthusiastic foreigners (F. JUNGHUHN for example) cooperated.

The first half of the nineteenth century was of importance for our geological knowledge of Indonesia as during that time much of the elementary information became known. The time for a synthesis had, however, not yet come. The only exception is the outstanding work of the botanist and geologist F. JUNGHUHN on Java (entitled "Java, deszelfs gedaante, bekleeding en inwendige structuur") which was published at Amsterdam in 1853.

After the foundation of the Department of Mining in the Netherlands East Indies in 1852, the geological investigations were in the hands of professional geologists. The publications of this Department and in particular the articles in the Annual Reports, testify to the high standard of their work. The outstanding personality during the end of the nineteenth and the beginning of the twentieth century was R. D. M. VERBEEK, but R. FENNEMA, J. A. HOOZE and C. J. VAN SCHELLE also distinguished themselves.

Geological surveys of vast areas in the Archipelago were carried out and comprehensive collections of fossils were described by specialists in Europe. In 1878 chairs of geology were endowed at the State Universities of Utrecht and Leiden, of which the first occupants were C. A. E. WICHMANN and K. MARTIN respectively. These scientific centres greatly furthered the geological investigations in the East Indian Archipelago.

During the first 25 years of the twentieth century not only VERBEEK, but also many other geologists and mining engineers carried out their fieldwork, both in and outside Java, from New

Guinea to Atjeh. H. A. BROUWER, L. J. C. VAN ES, C. W. A. P. 'T HOEN, M. KOPERBERG, A. TOBLER, N. WING EASTON, J. ZWIER-ZYCKI and numerous other investigators did admirable work during that period.

During the same period the many scientific expeditions were also of importance for our geological knowledge. In this connection may be mentioned C. E. A. WICHMANN's voyage to the eastern part of the Archipelago (1888–1889) and G. A. F. MOLENGRAAFF's investigations in Borneo (1893–1894); the Siboga expedition which was led by M. WEBER and advanced our knowledge of the sea floor a great deal; the military exploration of New Guinea, in which mining engineers and geologists took part, amongst others O. HELDRING, J. K. VAN GELDER and P. HUBRECHT; the German-Dutch frontier expedition with P. HUBRECHT; the Etna Bay expedition with P. MOERMAN; the New Guinea expeditions of 1903, 1907, 1909 and 1912 organized by the "Association for the promotion of scientific research in the Dutch Colonies" at Amsterdam, and the New Guinea expedition of 1920, organized by the "East Indian Committee for Scientific Research" at Batavia; H. WITKAMP's explorations in Sumba (1910), western Borneo (1923–1924) and eastern Borneo (1925); K. MARTIN's journeys for collecting molluscs in Java (1910); the expeditions of J. WANNER to Timor, Misoöl, Obi, Halmaheira (1909) and Timor (1911), followed by those of G. A. F. MOLENGRAAFF H. A. BROUWER, F. WECKHERLIN DE MAREZ OYENS to eastern Dutch Timor (1911) and of H. G. JONKER, H. A. BROUWER and H. BURCK to the same area (1916); P. and F. SARASIN worked for several years in Celebes (1893–1903) and in later years this island was investigated by ABENDANON (1909–1910); L. M. R. RUTTEN and W. HOTZ worked in Ceram and Buru (1917–1920); G. HENNY in Buru; O. POSTHUMUS and J. ZWIERZYCKI were members of the palaeobotanical Djambi expedition (1925).

All these expeditions, except those of P. and F. SARASIN and J. WANNER, were organized by Netherlanders. During the same period Indonesia was visited by several German investigators, e.g. W. VOLZ (northern Sumatra), K. DENINGER (Ceram and Buru), G. BOEHM (Sula islands and Misoöl), J. ELBERT (eastern part of the archipelago) and by the Selenka expedition to Trinil in Java which re-investigated the localities where E. DUBOIS found his vertebrates.

Finally, in the beginning of the twentieth century much important scientific exploration work was done by oil and private mining companies. The results of this work have in general not been published; however, to some of the geologists engaged by private mining companies, in this period, we owe several important publications. We may in this connection mention the names of J. AHLBURG, B. G. ESCHER, H. HIRSCHI, W. HOTZ, W. C. KLEIN, G. A. F. MOLENGRAAFF, A. TOBLER, J. WANNER, and FR. WEBER.

At the end of the first quarter of the twentieth century an enormous body of data was therefore available, as well as a number of tentative syntheses.

This period of geological research was ended by two books which summarized the existing knowledge, namely H. A. BROUWER's "Geology of the Netherlands East Indies" (1925) and L. M. R. RUTTEN's "Voordrachten over de geologie van Nederlandsch Oost Indië" (Lectures on the geology of the Netherlands East Indies) (1927). The latter book in particular gives a very complete survey of the development of our geological knowledge of Indonesia up to 1926. Much compilation was also done by the Department of Mining for its large scale geological maps (scale 1 ; 1,000,000), twelve sheets of which were published, together with an explanation, in the Annual Reports between 1915 and 1930.

The following period of 25 years was much more eventful than the previous one. At first there was a great expansion of the geological investigations and a great increase in the publications of the Department of Mining, but later a sharp decrease occurred because of the economical crisis in the early thirties. Still later investigation and publication came nearly to a complete stop, because of the war in the southwest Pacific and the political unrest in Indonesia in the post-war period.

In 1922 a Geological Survey was formed within the Department of Mining. It began its work with a systematic remapping of Sumatra and Java and organized extensive geological investigations in Borneo, Celebes, Buton, etc. During the Fourth Pacific Science Congress in Java in 1929 the new geological museum in Bandung was opened. The director at that time was A. C. DE JONGH and the staff comprised 37 university-trained scientists. This number was greatly reduced in later years because of the financial crisis and at the beginning of the second world war, when W. C. BENSCHOP KOOLHOVEN was director, only 17 were left.

During the war many mining engineers and geologists employed by the government and the private mining and oil companies, lost their lives. The geological museum at Bandung with its important collections and records suffered only a little.

Between 1925 and the beginning of the second world war much geological work was also done by private companies, but little of the work done by or on behalf of private mining and oil companies has been published. This is regrettable from the scientific point of view.

Apart from this work directed towards commercial ends, many voyages and expeditions were made during this period with only purely scientific ends in view, and their results have been published in detail. In this category belong the Snellius expedition in the eastern part of the Archipelago in 1929–1930 of which expedition the geologist PH. H. KUENEN was a member; the gravity measurements at sea by F. A. VENING MEINESZ on board the Dutch submarine K XVIII in 1929 and 1930; the Central Celebes expedition headed by H. A. BROUWER of which the government geologists W. H. HETZEL and H. E. G. STRAETER were members: the expedition to the Carstensz peaks of the Snow Mountain Range in New Guinea in 1936 under the leadership of A. H. COLIJN, in which the B.P.M. geologist J. J. DOZY took part; the expedition to the Lesser Sunda Islands by H. A. BROUWER and a number of pupils in 1937; the expedition of the Royal Netherlands Geographical Society to the Wissel lakes in New Guinea in 1939 headed by LE ROUX, of which R. IJZERMAN was the geologist.

In 1948 a general survey of the scientific work, which had been done in the Netherlands for the colonies in the 25 years after the first world war, was published at Amsterdam [1]. This work contains summaries by L. F. DE BEAUFORT on palaeontology, zoogeography and zoology, by L. M. R. RUTTEN on geology and by F. A. VENING MEINESZ on geodesy and geophysics.

The bibliography of the East Indian Archipelago and the adjacent areas, published by the Netherlands Geological and Mining Society and edited by R. D. M. VERBEEK, N. J. WING EASTON and J. F. STEENHUIS, contained, by 1949, more than 6700 works.

When we consider the geological work done during the last

---

[1] Report on Scientific work done in the Netherlands on behalf of the Dutch overseas territories (1918–1943). Published by Noord Hollandsche Uitgevers Maatschappij, Amsterdam, 1948.

century in Indonesia, it becomes clear that our knowledge is not less complete than that of other parts of the earth, apart from the regions within the sphere of western culture, where the geological sciences originated. It is true that our knowledge of vast areas is still rather fragmentary, but on the other hand, it is also true that most of the islands have been fairly well investigated and that in general the major trends of the structure and the geological development are known.

Indonesia in the last decades has become an area, which demonstrates the laws and mutual interaction of the processes leading to the formation of rocks, and the deformation of the earth's crust. These processes, called petrogenesis and tectogenesis respectively, are accompanied by active volcanism, seismic unrest and abnormal values of gravity. Thanks to the work of hundreds of geologists, who collected and published the fundamental observations, this region has become a testing ground for our conceptions of the evolution of the earth's crust.

In 1946 the present author was commissioned by the Netherlands East Indian Government to write a synthesis of all the earlier work. This work was published in two volumes towards the end of 1949 as a special publication of the Department of Mining in Indonesia by the State Printing Office at the Hague, bearing the title of "The Geology of Indonesia".

On 17 March 1951, TH. H. F. KLOMPÉ became the first occupant of the chair of geology at Bandung, endowed by the University of Indonesia. It is hoped that this new scientific centre in Indonesia will continue and expand the valuable work already done.

Physiography
When we are dealing with the geology of the East Indian Archipelago, we shall not heed the ever-changing political boundaries but only consider the natural geographical and geological conditions. We are dealing with the vast island empire between the Asiatic continent to the north-west, Australia to the south-east, the Indian Ocean to the south-west and the Pacific to the north-east. In this area two great mountain systems meet, viz. the more or less east—west Alpine system, which stretches from Gibraltar to Banda, and the Circum-Pacific mountains and island arcs. These two mountain systems join each other in the Archipelago and it is not surprising that they give rise to a most complicated physiographic pattern.

41

When we are studying this pattern, we can start from the major features of the relief and zonal structure of the area. There are two more or less stable regions which have been peneplained and were partly submerged in the most recent geological past, when sealevel rose approximately 100 metres relative to the land. These two stable regions are firstly the Sunda shelf with the Malayan peninsula and western Borneo in the western part of the Archipelago, and secondly the Sahul or Arafura shelf in the south-east. Parts of these regions are recently submerged borderlands of Asia and Australia. From the relief map (Fig. 3) it may be seen that on the floors of these shallow seas old river courses, dating from the Ice Age (Pleistocene), have been eroded.

The relief is much more pronounced in those parts where Pleistocene mountain building has taken place and at present is still taking place. It is here that the zonal structure, which had already in the eighteenth century been recognized by von RICHTHOFEN, is most clearly expressed in the morphology.

We can distinguish two principal systems which meet in the East Indian Archipelago, the Sunda system and the Circum- Pacific system. In the Indonesian region the latter consists of two main units, namely the system of the East Asiatic island arcs and the Circum Australian system.

The Sunda system belongs to the system of Alpine mountains and has a length of 7,000 kilometres. It extends from the Arakan Yoma chain in Burma, through the Andaman and Nicobar Islands, Sumatra, Java and the Lesser Sunda Islands to the Banda island arcs in the Moluccas. It is one of the world's longest continuous mountain systems, comparable in length with the Cordilleras de los Andes in South America. Along the entire length of this system, which is convex towards the south-west, two parallel belts of mountain chains, island arcs or submarine ridges can be distinguished. The belt at the inner, the concave, side is volcanic and the belt at the convex side is non-volcanic, but characterized by overfolding and overthrusting in an outward direction, as well as by negative isostatic anomalies.

The system of the East Asiatic island arcs, i.e. the north-western part of the Circum-Pacific system, starts in Kamchatka, in the north, continues across the Kurile Islands, Japan, the Riu-Kiu Islands and the Philippines to northern Borneo and Celebes. The arc of the

Philippines bifurcates into two elements in a southerly direction, of which the principal or Luzon arc continues to northern Borneo by way of Palawan and the Sulu Archipelago. Near Samar another branch leaves the principal one and passes through eastern Mindanao and the Sangihe Archipelago to Celebes (Sulawesi).

The Circum-Australian system extends from New Zealand in the south-east across New Caledonia to New Guinea (Irian). The northern part of this Circum-Australian system consists of two elements. The principal element forms the backbone of New Guinea, which extends from the "tail" of the island in the east as far as the Charles Louis Mountains in the west, where it is cut off by the structural elements of the Banda arcs, which belong to the Sunda Mountain system. The other element extends from the Bismarck Archipelago in the east along the northern coastal ranges of Irian and the Vogelkop to Halmaheira in the west.

The systems of young orogenesis mentioned above meet in the neighbourhood of the Sulu and Banggai islands, which form, geographically speaking, the boundary between the northern and the southern Moluccas and which are tectonically also in a special position.

II. STRATIGRAPHY

The geological formations
It is not surprising, that an area, which at present shows such a variety in topography, also had an eventful geological past. The distribution of land and sea, and of high and low areas, changed continually in the course of the geological history. Tracts of the earth's crust, which at present form deep submarine basins, may have been high mountain chains in the past; and conversely, the present-day mountains have been zones of continuous subsidence for tens of millions of years.

The areas which were lifted above sea level, were slowly attacked and worn down by erosion and consequently deeper and deeper parts of the earth's crust were exposed. Sediments which recrystallized in the depths of the earth and holocrystalline igneous rocks appeared at the surface after prolonged erosion had taken place. The insoluble detritus of these regions was transported to the adjacent marine basins as clay or sand, or as pebbles, if the differences in relief were particularly pronounced. In these seas organisms which abstract calcium carbonate from sea water such as corals, molluscs and foraminifera were living, and consequently the marine clays often contain calcium carbonate (marls), or show intercalations of limestone beds, formed by coral reefs etc. When in the neighbourhood active volcanoes were present, we also find intercalations of tuffs, volcanic breccias and lava flows. Occasionally, when low, marshy coasts were present, fresh water sediments were formed. In these fresh water sediments layers of peat may occur which are gradually transformed into lignite and finally into coal and anthracite.

In this fashion a great variety of geological formations originated, variable in composition, thickness and distribution. From the nature of such a formation, its facies, the geologist can reconstruct the environment in which the deposition took place [1]. He can state whether we are dealing with a continental or a marine deposit; whether it was formed near the coast or in the deeper sea and far from the land; whether the differences in relief were great (coarse sediments such as conglomerates) or small (clear water or supply of fine sediment); whether earth movements occurred during the deposition or whether the conditions of erosion and deposition remained practically unchanged; and whether there was any volcanic activity.

By studying the chemical and mineralogical composition of the rocks the geologist is often able to say what happened to these rocks after their formation. In sediments mineralogical changes are caused by increases in temperature and pressure (under a cover of younger deposits) and/or by tectonic deformations. These changes are called metamorphism. Metamorphism is often accompanied by addition and/or abstraction of chemical elements and it can sometimes lead to a general change, ultra-metamorphism or transformation, of the sediments. When this has happened, the sedimentary rocks resemble the rocks which crystallize from magmas, and the traces of their

[1] Abyssal facies   = deep sea of more than 1,000 metres.
  Bathyal facies   = deep sea from 1,000 to 200 metres.
  Neritic facies   = shallow sea up to 200 metres.
  Littoral facies   = beach formations.
  Paralic facies   = alternating sea and land deposits.
  Terrestrial facies = land deposits.
  Lagoonal facies   = brackish water deposits.
  Volcanic facies   = deposits which are entirely or partly composed of volcanic materials, such as tuffs, lahar breccias and lava flows.

44

origin have for the greater part been obliterated. On the other hand it may happen that igneous rocks which originally possessed a granular crystalline structure, acquire a schistose one when subjected to pressure, and resemble metamorphic sedimentary rocks.

The attitude of the sedimentary strata also provides important information on their origin and history. We know that sediments tend to be deposited in horizontal layers. After their deposition, stresses originate in the crust and deformations, called tectonic deformations, occur, giving rise to folds, faults and even overthrusts of one series upon the other.

From the study of the distribution, the chemical and mineralogical composition, the facies, the attitude and the mutual relations of the rocks, the geologist can reconstruct the conditions which existed in and on the earth. He can trace the former distribution of land and sea and make some general statements on the topography, the depth of the seas and the distribution of active volcanoes. A first requirement for such reconstructions of former conditions is that the age of the sediments should be accurately known, as an inaccurate correlation of these sediments would give an entirely unreliable picture of former conditions.

The study of geological formations, their distribution, succession and mode of origin, as well as the determination of their relative ages is called stratigraphy. The methods used are in the first place direct observation in the field where characteristic beds are traced from one exposure to the other and in the second place the age determination of the fossils in the sediments.

The determination of the absolute age of rocks is nowadays also possible in favourable cases. This is done by using the disintegration of radio-active elements which produce helium and certain lead isotopes. Thus an absolute time scale for the geological formations has been obtained. One must not expect these results to be accurate within, say, one million years, but it does give the order of magnitude of the geological periods.

In Table 1 the international stratigraphic column is shown, as well as the length and the age of the principal periods, according to A. HOLMES. For the younger geological periods we shall require a more detailed sub-division than is shown in this general stratigraphic table.

The fossiliferous part of the geological evolution, which lasted more than five hundred million years, was preceded by a much

45

longer non-fossiliferous period [1], which according to A. HOLMES, goes back more than three thousand million years, when our Earth and the other planets originated and took their places in orbits round their mother star, the sun.

TABLE 1

International stratigraphic table of the fossiliferous part of the geological formations

Group	Period		Length of period in millions of years	Age of its base in millions of years
Cainozoic	Quaternary		$\frac{1}{2}$–1	$\frac{1}{2}$–1
	Pliocene	Neogene	11	12
	Miocene		14	26
	Oligocene	Palaeogene	12	38
	Eocene		20	58
Mesozoic	Cretaceous		69	127
	Jurassic		25	152
	Triassic		30	182
Palaeozoic	Permian		21	203
	Carboniferous		52	255
	Devonian		68	323
	Silurian		27	350
	Ordovician		80	430
	Cambrian		80	510
Archaic			circ. 2500 –3000	

Palaeozoic

The oldest fossiliferous deposits of Indonesia belong to the Middle and Upper Palaeozoic. Examples are: the Silurian and Devonian formations of the Snow Mountain Range in New Guinea, the Devonian formation of the Telen area in Borneo and the Carboniferous formation of the Barisan Mountains in Sumatra. This, however,

[1] The fact that pre-Cambrian formations yield no or only a few problematic fossils, does not mean that life on earth only started in Cambrian times. The first living organisms were very primitive and only consisted of protoplasm. Later a fairly great variety of species and genera developed from these primitive ancestors, but they did not yet possess parts which could be preserved in sediments. In Cambrian times evolution had progressed so much, that hard parts of a great number of species could be preserved and are now found as fossils.

46

does not mean that the geological evolution started in these times. Earlier, in Lower and pre-Palaeozoic times, orogenesis and volcanism had produced a crust of crystalline schists, gneisses and holocrystalline igneous rocks (granites, etc.). The existence of this older crust is proved, for example, by the presence of granites in Sumatra formerly belonging to mountain chains, on which fossiliferous Carboniferous sediments were deposited. In the Schwaner Mountains of central Borneo, erosion products of crystalline schists and gneisses are found in deposits of Permo-Carboniferous age. As far as Indonesia is concerned, nothing more is known about the evolution of this older crust. In the adjacent regions of south-eastern Asia (Cambodja and Australia), fossils of Cambrian age have been found. In Australia it is even possible to distinguish between Lower, Middle and Upper pre-Cambrian formations.

As regards Indonesia, we can only state that the previous evolution of the crust had already produced an extensive land area in Early Palaeozoic times. This area was presumably part of the Gondwana continent of the southern hemisphere, to which Australia also belonged. The land area which in Early Palaeozoic times was situated between Asia and Australia is called the Indonesian Primeval Continent by the author [1].

This primeval continent forms the so called crystalline basement, and is made up of crystalline schists, gneisses and plutonites. The rocks of this crystalline basement can mostly be easily distinguished from the younger deposits, which have suffered little or no recrystallization and often contain fossils. We call these younger deposits the sedimentary skin or epidermis of the earth's crust.

Formerly it was thought that crystalline basements were invariably of very great age, namely Archaic. Nowadays it is known that this is not the case. During orogenesis, sediments at a certain depth in the earth are subjected to all kinds of mineralogical and chemical changes. These metamorphosed sediments may later be

---

[1] This opinion is corroborated by a recent publication of R. W. FAIRBRIDGE ("The Sahul Shelf, northern Australia; its structure and geological relationships". J. Roy. Soc. Western Australia, Vol. 37, 1953, 33 p.). Stratigraphic and palaeographic data indicate that much of Northern Australia and its continental shelf region have been continental since Archaean times. Palaeographic and faunistic connections suggest that this continental area (either land or as shelf) formerly extended far to the northwest. The Wegenerian idea of an Australian continent, drifting several thousand kilometres to impinge upon the Moluccan region in late Mesozoic to Tertiary time, cannot therefore be entertained.

exposed by erosion. The only thing that can be said with certainty on the age of the basement, is that it must be older than the first non- or less metamorphosed sediment which was deposited on top of it.

From Silurian times New Guinea belonged to a geosynclinal subsiding tract, the Papuan geosyncline, which stretched along the edge of Australia and through the eastern part of that continent as far as Tasmania.

The subsidence in the Papuan geosyncline continued (with some interruptions) into Cainozoic times. During the very latest part of the earth's history, this changed into a rising movement, which locally even lifted the mountain peaks above the snowline in the tropics (See figure 4).

The presence of marine Silurian sediments in the Snow Mountain Range of New Guinea is proved by the occurrence of pebbles containing a typical index fossil for the Silurian (Hallysites wallichi Reed).

In the western part of Indonesia, the oldest marine deposits dated by means of fossils, belong to the Lower Devonian (acc. to the corals Heliolites porosus GOLDFUSS and Clathrodictyon cf spatiosum BOEHNKE, described by M. G. RUTTEN from sediments of the Danau formation in the Telen region in eastern central Borneo). In Devonian times the subsidence was presumably still restricted to a narrow strip, which extended across the old land area from Borneo to south-eastern Asia. The part of this subsiding strip which is situated in the Sunda region, is called the Anambas geosyncline, after the islands of the same name on the Sunda Shelf between Borneo and the Malayan peninsula.

Much more is known about the distribution of the younger Palaeozoic (Carboniferous and Permian). In various islands fossil floras and faunas, belonging to this period, have been found. Sometimes it is not possible to distinguish between the Fusulinas (which are used as index fossils) of Permian and Carboniferous age. These formations are therefore sometimes taken together as the Permo-Carboniferous.

It is a striking fact, that the fossils found in the Palaeozoic of Indonesia, mostly occur in sediments which were deposited in shallow seas not far from the land. It seems probable, that in Late Palaeozoic times the old land area broke up into a number of islands

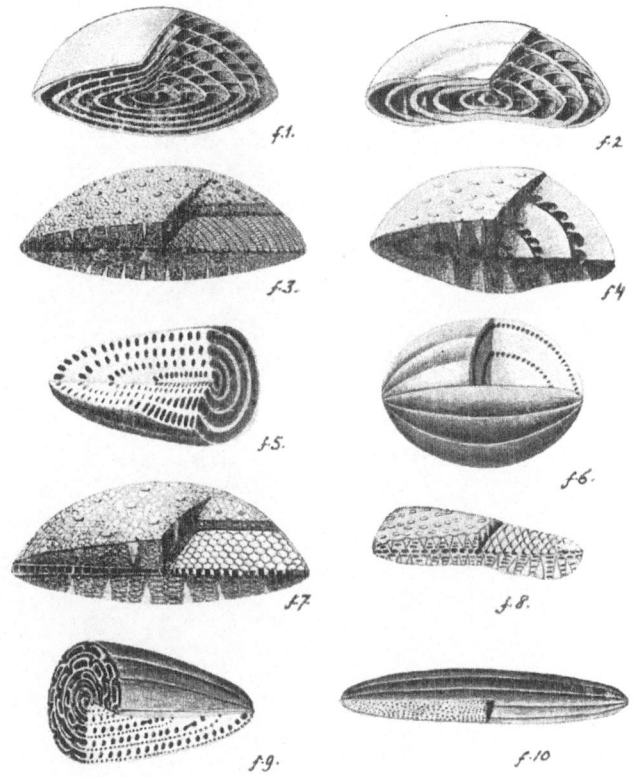

Fig. 6 Tertiary foraminifera (acc. to *Umbgrove*).

f1 Nummulites, Lower Tertiary.
f2. Assilina, Lower Tertiary.
f3. Discocyclina, Lower Tertiary.
f4. Pellatispira, Lower Tertiary.
f5. Alveolina, Lower Tertiary.
f6. Flosculina, Lower Tertiary.
f7. Lepidocyclina, Middle Tertiary.
f8. Myogypsina, Middle Tertiary.
f9. Alveolinella bontangensis, Middle Tertiary.
f10. Alveolinella (recent type), Middle Tertiary till recent.

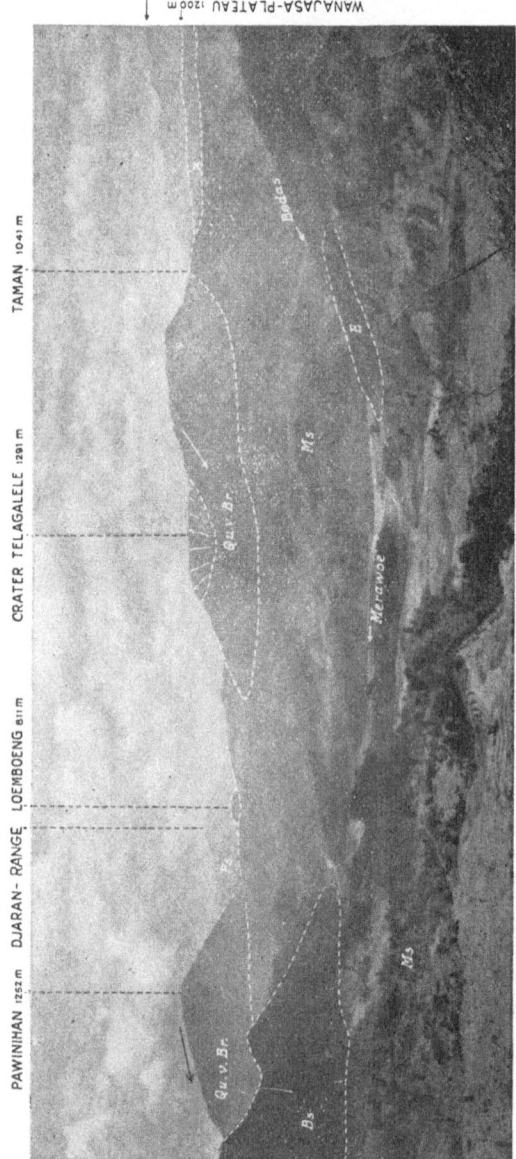

Fig. 24 Panoramic view of the Karangkobar region with the sliding volcanoes Pawinihan and Telagalele, as seen from Mt. Pining at the WNW.

and larger land masses, separated by shallow seas. Only locally, in the Permo-Carboniferous of western Borneo, indications of Late Palaeozoic deep sea sediments are found. These sediments originated in a comparatively narrow strip, 800 kilometres long, which extended across Borneo, from Kuching in the west to Berau in the east. In a southerly direction the abyssal and bathyal sediments of the "Danau formation" pass into neritic and paralic sediments. This change occurs within a distance of some tens of kilometres, and proves that already in Permo-Carboniferous times, ridges and troughs, called geanticlines and geosynclines, existed side by side. Similar contrasts of high and low characterized the relief in Indonesia during the entire further development in Mesozoic and Cainozoic times.

The distribution of the flora in Carboniferous times is also of interest. In the southern hemisphere, the Gondwana continent, the Glossopteris flora is found, whereas eastern Asia in Early Carboniferous times was characterized by the Gigantopteris flora. The northern flora occurs in the Djambi nappes of Sumatra, in Malaya and in Borneo. In the southern slope of the Snow Mountain Range in New Guinea, a flora has been found which combines elements of the northern flora (species of Pecopteris and Taeniopteris) with Vertebraria, a typical element of the southern flora. In Carboniferous times mixing of northern and southern floras of land plants was therefore possible. The distance between the land areas was apparently not large enough to prevent this (see note on p. 47).

The breaking up of the old land area in Late Palaeozoic times, was locally associated with the effusion of basaltic lava flows on the floors of the subsiding tracts which were coming into existence. In this category belong the basalts, trachybasalts, alkali-trachytes and -rhyolites in the Permian of the Sonnebait nappe in Timor. This group of volcanic rocks is, chemically speaking, related to the eruption products of the Indian and Atlantic Ocean, which are grouped together as the Atlantic suite of igneous rocks. We shall return to this subject when we are discussing the petrographic provinces in the chapter on volcanology.

Mesozoic
In the Mesozoic Era the Eurasian land masses of the northern hemisphere and the disintegrated Gondwana continent in the

southern hemisphere, became separated by a wide belt of seas and islands. This belt has been given the name Tethys by geologists; its seas formed a continuous connection between Europe to the west and Indonesia to the east. This explains the remarkable resemblance between the fossils of Europe and Indonesia in this era. This, of course, is of great importance for the correlation of the geological formations in Indonesia with the standard stratigraphy of Europe. In Mesozoic times the Tethys sea formed a single extensive faunal province. Only as late as Eocene times were the sea connections between the Mediterranean and the Archipelago finally interrupted by orogenic changes. The Tethys was divided into a number of isolated seas, separated by extensive land areas. In these isolated seas life evolved independently. The Indo-Pacific faunal province, characterized by its own types of organisms, originated only after the Mesozoic.

The subsidence of parts of the Indonesian Primeval Continent in Late Palaeozoic times was followed by a long and complicated cycle of mountain building, a process which persists into the present. This mountain building, orogenesis, determines the major features of the present structure of Indonesia. The geological history of Indonesia is the history of this orogenesis and the processes associated with it. It will be clear that the processes of mountain building, causing as they do, large vertical movements, both up- and downwards, are of decisive importance for the distribution and nature of the sediments. We shall therefore add a few general remarks on the relations between orogenesis and sedimentation.

The effect of orogenesis is that mountain chains, geanticlines, are lifted up. These ridges are volumetrically compensated by the subsidence of the adjacent regions. A mountain chain is therefore on one or either side accompanied by subsiding tracts, which are called geosynclines.

The width of these "waves" in the earth's crust, originating during the orogenesis, is approximately 100–200 kilometres. The adjacent narrow subsiding tracts which compensate the uplift of the geanticlines, are called geosynclines, but they should not be confused with the much wider belts within which the process of mountain building as a whole takes place, and which are also called geosynclines. The Tethys sea, which has already been mentioned, belongs to this latter category. These geosynclinal areas have a width of more than 1.000 kilometres. They are therefore of a much greater order of magnitude than the narrow geosynclinal troughs which are formed during orogenesis.

The narrow geosynclinal zones of subsidence can form deep basins, when the rate of supply of sediments and products of volcanic eruptions is small compared with the rate of subsidence. This is illustrated by the deep east of the Philippines, which has a depth of about 10,000 metres and the deep south of Java with a depth of 7,450 metres. In such deeps, sediments of an abyssal or bathyal facies may be formed. In many cases, however, the supply of sediments is large enough to keep up with the subsidence. In that case shallow seas with adjacent low coastal tracts originate in which thick series of neritic, littoral, or fresh water facies alternate with each other, often attaining a total thickness of several thousands of metres. The original surface of the land may lie at several thousands of metres below the present sea level at the end of such a sedimentary cycle. Yet it does not follow, that at any time the sediments were laid down in a bathyal or abyssal environment.

The differential vertical movements of the crust are sometimes accompanied by external volcanism; for example, by the effusion of lava on the floors of the subsiding tracts or by volcanoes on the geanticlinal ridges.

These parallel geanticlinal and geosynclinal zones are characteristic of regions where orogenesis is taking place, as it has been in Indonesia since Devonian times. It will be clear that orogenesis causes great contrasts in relief, and that consequently the environment and composition of the sediments change over comparatively short distances.

In this book a summary of the distribution of the various Palaeozoic and Mesozoic formations and of the fossils by which they are characterized will not be given [1]. In Chapter VI we shall deal only with the evolution of the Sunda region in greater detail. This will demonstrate the close relations between orogenesis and the distribution and facies of the sediments.

Cainozoic
The rocks of the crystalline basement, and the Late Palaeozoic and

[1] For details about the stratigraphy and paleontology of the Mesozoic in Indonesia the reader is referred to "The Geology of Indonesia", Vol. I A.

Mesozoic sediments, are of great importance for the reconstruction of the older geological history of Indonesia. Its present appearance, however, was mainly determined by the events in the more recent part of its geological history. The Cainozoic deposits, i.e. those of the Tertiary and Quaternary taken together, form more than three quarters of the direct sub-soil of this vast area, and often attain enormous thicknesses.

In many places the Tertiary sediments are separated by an unconformity from the underlying basement, which is Mesozoic or older. An unconformity is a break in the deposition, a number of pages are absent, as it were, from the book of stone in which the geologist reads the local geological history.

A subsiding zone, in which sediments have been deposited, may have been lifted above sea level and subjected to denudation, instead of sedimentation. When such a zone is submerged again later on, deposition continues, but there is a break in the record of its history (see Fig. 5).

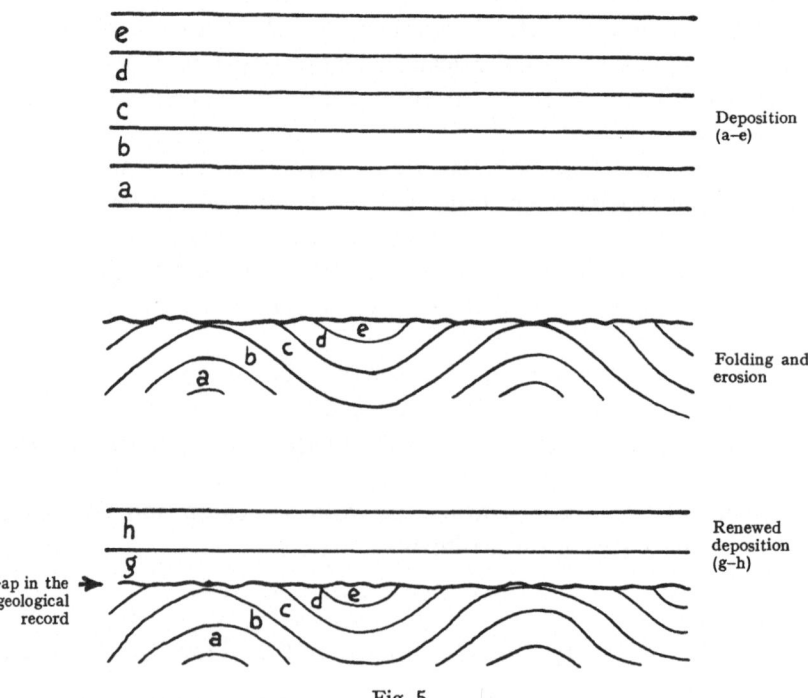

Fig. 5

Sketches showing how an angular unconformity comes into existence.

If the uplift and the following subsidence are purely vertical, without bending or deformation of the deposits, the newer and older series are parallel. In this case we have a series of parallel beds, which contain a disconformity. If, however, folding or dislocation of the older series has occurred, the younger series is separated from the older one by an angular unconformity. Most unconformities which represent important breaks in the geological history, are angular unconformities. It often happens, that during this break in the sedimentary record volcanic rocks intrude the older series. In that case these volcanic rocks are lacking in the younger sediments, which were deposited after the break. The volcanic activity can be dated as having taken place after the deposition of the youngest bed before the interruption, and before the formation of the oldest deposit after the interruption.

The Tertiary sediments are in general separated from the basements by a clear angular unconformity. A few exceptions are known, for example in Ceram where the Cretaceous is conformably overlaid by Eocene beds, both of bathyal facies.

In the same island the Eocene also occurs with a littoral facies. This type of Eocene is separated from the Mesozoic by a clear unconformity. In Halmaheira Cretaceous and Eocene seem to be conformable.

The Tertiary sediments show a great variety of facies types. The abyssal facies, however, as for example the red clay of the open oceans, has not been identified with certainty. It goes without saying, that in the deeps of Indonesia, deep-sea deposits are at present being formed — the results of the Snellius expedition confirm this — but it seems that the younger orogenic movements have not yet raised typical abyssal deposits of Tertiary age above sea level.

The Tertiary deposits in many areas are characterized by a strong volcanic activity, the products of which are either formed locally or transported through the air and by water from adjacent volcanic zones to the basins, where sedimentation takes place.

Subsiding troughs, situated at the edge of more stable areas (as for example the Sunda region) are typical of Tertiary times in Indonesia. These troughs were elongated in shape and separated by partly submarine sills. In their central parts, sedimentary series of great thickness were deposited in a comparatively short time. The

Tertiary sediments in eastern Atjeh reach a thickness of 8 kilometres, in Palembang 6 kilometres and in south-east Borneo even thicknesses of 12 kilometres. UMBGROVE has given the name of idiogeosynclines to these subsiding troughs. They are of great importance for the formation of the mineral deposits of Indonesia, c.q. petroleum, lignite and coal.

This is the reason why so much attention has been given to the stratigraphic study of these basins. Various methods have been used, first of all those based on the fossil content using either macrofossils, such as molluscs, corals and vertebrates, or microfossils such as foraminifera, ostracods and diatoms, and secondly methods using the mineralogical composition of the sediments, the water content of the lignites, etc.

The molluscs in the Indo-Pacific province, developed independently of the European fauna after Mesozoic times. It is therefore not easy to determine the position of the Cainozoic strata in Indonesia in the international stratigraphic table, which is based on European species. Professor K. MARTIN has therefore used the so-called percentage method. MARTIN determined what percentage of the mollusc species in a fossil fauna belonged to living species. The older the fauna, the smaller is this percentage. He used the following figures:

Quaternary	70%	living species
Pliocene	50–70%	,, ,,
Upper Miocene	20–50%	,, ,,
Lower Miocene	8–20%	,, ,,
Eocene	0%	,, ,,

This is a statistical method, and it works better when a greater number of species in the fossil fauna is used. One definite disadvantage is that the recent fauna is not fully known. Littoral forms have been collected and described, but faunas from greater depths are only incompletely known. A second disadvantage is the lack of knowledge of the variation within a given species and the vagueness of the boundaries between the species, as well as between the subspecies. All this, of course, influences the percentage figures.

Many favourable results have been obtained by means of this method but some mistakes in the correlation of beds have occurred.

OOSTINGH has further elaborated the mollusc stratigraphy of Indonesia. This author has distinguished certain types of faunas and also tried to find certain species which could serve as index fossils.

Bantamian	Lower	Pleistocene
Sondian	Upper	Pliocene (Astian)
Cheribonian	Lower	Pliocene (Plaisancian + ? Pontian)
(Tjiodeng)	Upper	Miocene (= Sarmatian)
Preangerian	Middle	Miocene (= Vindobonian)
Rembangian	Lower	Miocene (= Burdigalian)
(West Progo)	Oligo-Miocene	(= Aquitanian)

In 1946 UMBGROVE proposed a statistical percentage method for the Cainozoic deposits based on corals. He gives the following percentages:

Pleistocene-Recent	70–100%	living species	
Pliocene (Tertiary h)	50– 70%	,,	,,
Upper Miocene (Tertiary g)	30– 50%	,,	,,
Middle Miocene (Tertiary f)	10– 30%	,,	,,
Lower Miocene (Tertiary e)	0– 10%	,,	,,
Oligocene (Tertiary c-d)	0– a few %		,,
Eocene (Tertiary a-b)	0%	,,	,,

REINHOLD in 1936–37 did the same for the diatoms and gave the following figures:

Lower Pleistocene	circ. 70%	living species	
Upper Pliocene	circ. 60%	,,	,,
Mio-Pliocene	circ. 50%	,,	,,
Middle Miocene.	circ. 30%	,,	,,

These methods and figures have, however, not yet been fully verified.

Foraminifera have been successfully used as a basis for correlation of sedimentary strata. The large oil companies all have micro-palae-ontologists in their service, who are specialists in this subject. Originally, only the large foraminifera which can be recognised by means of a hand lens, were used but in the last 20 years much attention has also been given to the microscopically small representatives of this group of animals. The great advantage of using foraminifera is that species from bore-hole samples can be used.

VAN DER VLERK and UMBGROVE established in 1927 a letter-classification of the Tertiary in Indonesia, based on large foraminifera. On purpose, the names of the international stratigraphic column were avoided. In later years it has become possible to carry out a correlation with European stratigraphy, in particular by means of the Alveolinellidae. On the other hand it appeared impos-

sible to confirm some of the sub-divisions used in the letter-classification. M. G. RUTTEN and VAN DER VLERK gave in "The Geology of
Indonesia" (1949, Table 13) the following stratigraphic table, based
on large foraminifera. See Fig. 6 and Table 2.

Tertiary g–h	= Pliocene (including Pontian)	
,,	g (in part)	= Upper Miocene (Sarmatian)
,,	$f_2 + {}_3$	= Middle Miocene (Vindobonian)
,,	$e_5 + f_1$	= Lower Miocene (Burdigalian)
,,	$e_1 - {}_4$	= Oligo-Miocene (Aquitanian)
,,	c + d	= Oligocene
,,	a + b	= Eocene

Index fossils for the Eocene are: Nummulites, Assilina, Pellatispira, Borelis (= Alveolina), Discocyclina. Characteristic for the
Oligocene is: Nummulites fichteli-intermedia; for the Aquitanian:

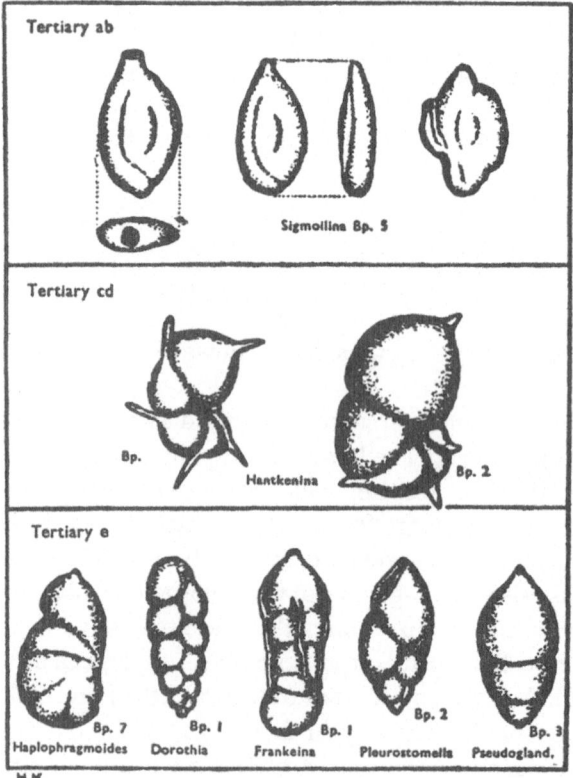

Fig. 7

Some smaller foraminifera (enlargement 15 ×). Index fossils for the Lower Tertiary
of SE Borneo acc. to MOHLER.

TABLE 2

Letter classification of the Tertiary in Indonesia based on the occurrence of large foraminifera. Acc. to new data of M. G. RUTTEN, and some additions of VAN DER VLERK.

TERTIARY

FOSSIL	a/b	c	d	e $_{1-4}$	e $_5$	f $_1$	f $_{2,3}$	g/h	Quaternary
Nummulites (= Camerina)									
N. javana = N. perforata									
N. djokdjokartae, N. vredenburgi									
N. fichteli-intermedia									
Assilina									
Pellatispira									
Biplanispira									
Heterostegina									
H. borneënsis									
Spiroclypeus									
S. vermicularis									
Other species of Spiroclypeus									
Cycloclypeus									
Approximate distribution according to TAN, 1932 { C. koolhoveni									
C. oppenoorthi									
C. eidae									
C. indopacificus									
Cycloclypeus (Katacycloclypeus)									
Borelis (= Alveolina)									
Borelis (Fasciolites)									
Borelis (Flosculina)									
Neoalveolinella pygmaea									
Flosculinella									
Alveolinella									
Austrotrillina (Trillina howchini)									
Lepidocyclina									
L. (Polylepidina)									
L. (Lepidocyclina)									
L. (Nephrolepidina)									
L. (Trybliolepidina)									
L. (Eulepidina)									
L. (Multilepidina)									
Miogypsina									
M. (Miogypsinoides)									
M. (Miogypsina)									
Discocyclina									
Approxim. % of living molluscan spec.				0%—8%	8%—35%			35%-60%	< 60%
Approximate correlation with Europe	Eocene	Oligocene		Miocene				Pliocene	Quaternary

57

Spiroclypeus; for the Burdigalian; Trillina howchini, Katacycloclypeus, Miogypsinoides; in the Vindobonian many Lepidocyclinae
and Miogypsinae occur, whereas in the Sarmatian these latter two
groups had already become extinct.

The small foraminifera, often less than 1 millimetre in diameter,
are characterized by a great number of species and a relatively small
number of individuals in the faunas (Fig. 7). Index fossils with a large
distribution are scarce. Standard collections are used for local
correlations of bore hole samples within a given basin.

Characteristic for the Eocene are: Globigerina eocena, for the
Eo-Oligocene: Hantkenina, for the Tertiary 'e' the species Haplo-
phragmoides, Dorothia, Frankeina, Pleurostomella and Pseudo-
glandulina. Orbulina universa only occurs after the Burdigalian.

In the last twenty years, vertebrates have become important for
the stratigraphy of the Upper Neogene and the Quaternary of Java,
thanks to work by VON KOENIGSWALD.

Vertebrate fauna according to VON KOENIGSWALD

Age	Fauna	Typical representatives
Holocene	Sampung	Recent species and some which at present are extinct in Java. Cervus eldi.
Upper Pleistocene . .	Ngandong	Cervus palaeojavanicus, Sus terhaari, Homo neanderthalensis-soloensis.
Middle Pleistocene . .	Trinil	Cervus (Axis) lydekkeri, Elephas cf. namadicus, Pithecanthropus erectus
Lower Pleistocene . .	Djetis	Cervus zwaani, Meganthropus and two kinds of Pithecanthropus
Upper Pliocene . . .	Kaliglagah	Mastodon bumiajuensis, Cervus stehlini.
Middle Pliocene . . .	Tjidjulang	Elephas planifrons, Merycopotamus, Hippopotamus.

The correlation with the vertebrate faunas of India has not yet
become possible. In Java and southern China Equus, which elsewhere
is thought to mark the boundary between Pliocene and Pleistocene, does not occur. VON KOENIGSWALD correlates the Tjidjulang
fauna of Java, which alternates with marine strata containing a

58

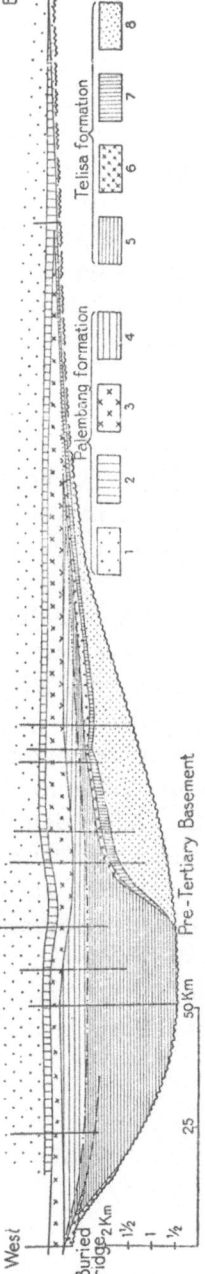

Fig. 8

Stratigraphic cross section through the southern part of the Neogene basin, South Sumatra, prior to Post-Pliocene folding. The vertical lines indicate the bore holes in which the successive beds are determined (acc. to DUFOUR, Proc. Third World Petroleum Congress, 1951).

1. Upper Palembang and upper part of Middle Palembang, Upper terrestrial facies

2. Lower part Middle Palembang, Upper paralic facies

3. Lower Palembang A, Upper large foraminifera facies

4. Lower Palembang B, Upper small foraminifera facies

5. Telisa Shale, Lower small foraminifera facies

6. Telisa (Baturadja) Limestone, Lower large foraminifera facies

7. Lower paralic facies

8. Telisa Sand, Lower terrestrial facies

Cheribonian mollusc fauna and therefore must be considered as of Early Pliocene age, with the Tatrot fauna of India, which contains Equus and therefore should belong to the Early Pleistocene.

It has been thought, that the water content of lignite could be used for stratigraphic ends. The water content of peat decreases with time and with increasing thickness of the sedimentary overburden. In very young brown coal the content of water, which is chemically bound to vegetable matter, exceeds 20%, but in Early Tertiary coal this has decreased to 3-7%. However, the water content does not depend so much on the age, as on the conditions of temperature and pressure to which the deposits have been subjected. It even happens that young brown coal has been metamorphosed on a large scale into real anthracite. The Bukit Assam coal mine in southern Sumatra produces Pliocene brown coal, which has thus been raised in rank.

Finally, it is possible to use the mineralogical composition for the local correlation of sediments within certain basins, a method which has been worked out by EDELMAN and DOEGLAS.

During the Cainozoic era, a particular fauna of species, which cannot directly be correlated with fossils from the classical stratigraphical localities, developed within the Indo-Pacific region. Moreover, as the sediments originated in the more or less isolated idio-geosynclinal basins, which developed independent of each other, a great variety in facies, thickness, intensity of folding and unconformities is found. (See fig. 8). The detailed stratigraphy of the Indonesian Cainozoic which has been established for most of the basins, and the correlation with the international stratigraphic column, must therefore be considered as an outstanding achievement of the stratigraphers and palaeontologists who, in the last decades, have worked in Indonesia.

We shall not deal with the Cainozoic formations and their distribution in any more detail. A few examples will be discussed in the chapter on the geological evolution.

III. VOLCANISM

Active Volcanism

We must distinguish between the present-day, active volcanism and the volcanic activity in the course of the geological history (=palaeo-volcanism). The volcanic activity as we observe it at present in Indonesia is no more than a still from a moving picture of the volcanic activity, which has taken place in Indonesia from the oldest times till now.

In Indonesia 128 centres of volcanic activity occur, of which 78 have erupted in historical times (i.e. since c. A.D. 1600), 29 are in the solfatara stage and about 21 are solfatara fields which are not obviously connected with a volcano. The number of volcanoes which are in a more or less advanced stage of exhaustion and disintegration is much greater and exceeds 500.

Indonesia is at present the most volcanic region of the earth. In Japan, which comes second, 82 volcanic centres are known of which 55 have been active in historic times.

The number of volcanoes, which between 1920 and 1941 showed eruptive or increased activity, averaged annually just over 10. Volcanic catastrophes causing loss of life and/or damage to cattle and crops occurred since 1800 on the average once in every three years. The greatest disasters were caused by the eruption of the Tambora volcano in 1815, when 92.000 people were killed and by the eruption of Krakatau in 1883, when more than 36.000 people perished. The effects of such catastrophes can be reduced to a minimum by continuous scientific observation and timely warnings to the population.

In general, however, it is true that the volcanoes are the benefactors of Indonesia, because they make the soil fertile by their showers of ash. It is only because of these volcanoes that Java can

61

support a population of 50 million (on the average more than 400 people per square kilometre), whereas Borneo, which does not possess volcanoes, has a density of population of only 4 per square kilometre.

The map (Fig. 9) shows the arrangement of the volcanoes in a number of zones. These zones form in general the inner arcs of mountain systems; their outer arcs are non-volcanic. The Sunda Mountain System is a typical example. Its volcanoes are situated on the axis of a large geanticline which stretches from the Barisan Mountains in Sumatra through Java to the Lesser Sunda Islands. The non-volcanic outer arc runs through the islands west of Sumatra and the submarine ridge which cuts the deep south of Java into two. In the northern Moluccas, two zones of active volcanism the Mina-hassa and the Halmaheira zone, are found, whose convex sides face each other. These volcanic zones have a non-volcanic outer arc in common which has the shape of a complicated and for the greater part submarine rise in the floor of the North Moluccan Sea.

Palaeo-volcanism
The distribution of active volcanism points therefore to a definite connection with mountain structures. Not only the geographical distribution is important, but also the factor time. Only by study-ing the relations between volcanism and orogenesis in space and time the general principles which characterize the geological evolution of Indonesia can be understood.

The only external expressions of active volcanism are the spec-tacular eruptions, or the quiet outflow of volcanic gases in solfataras, fumaroles and mofettes. Moreover, we only observe the volcanic activity on the land and in shallow seas, but it is not known whether at present lava flows are formed on the floors of deep troughs. The many thousands of metres of water, which hide the floors of these deeps from direct observation, form effective buffers which conceal these eruptions. Nor does the present day volcanic activity teach us anything about the volcanic processes which occur within the earth. If, however, we study the igneous rocks in mountain struc-tures of various ages, it appears that a relation exists between the internal and the external volcanism, and between volcanism in its wider sense and the tectonic deformations of the earth's crust.

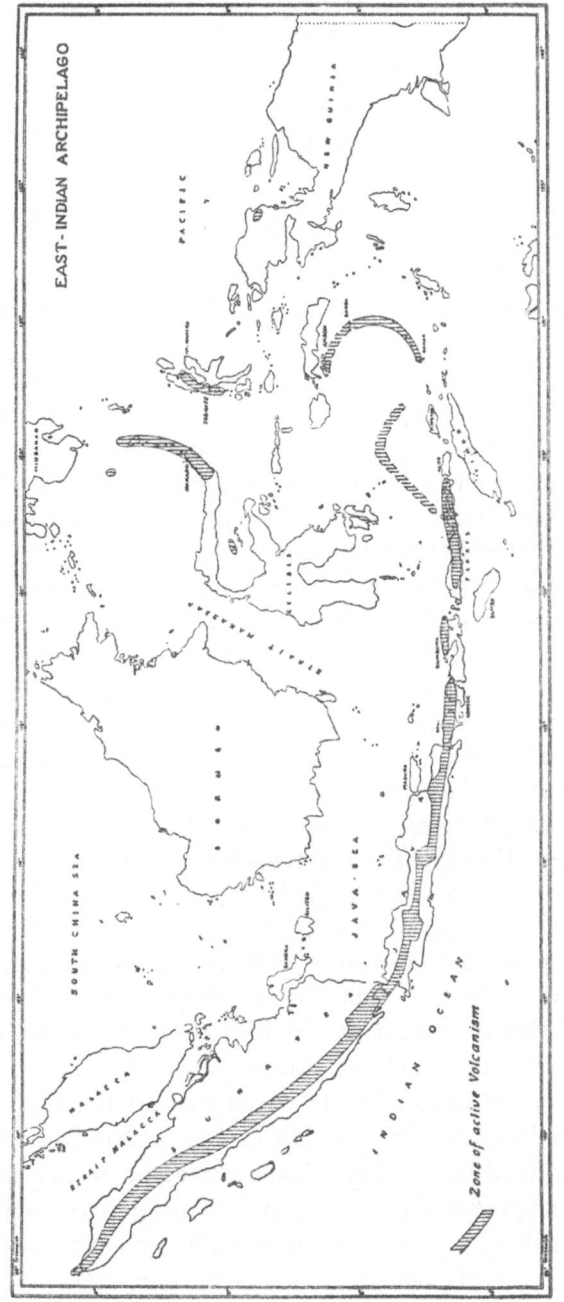

Fig. 9

Outline map of Indonesia, showing the zones of active volcanism.

63

Geosynclinal deposits which have been lifted above sea level by later orogenesis, and are accessible to direct observation, are often accompanied by sodium-rich basalts (spilites), masses of serpentine and other intrusive and extrusive rocks poor in silica. This is a series of igneous rocks, which are closely related from the chemical and mineralogical point of view, and which have been given the name of ophiolites. Judging from the volume of the eruption products, the volcanic activity in geosyclines, in many cases, cannot have been less important than the volcanic activity on geanticlines, as we know it from Sumatra, Java and the Lesser Sunda Islands.

The volcanic activity on geanticlinal ridges is, however, much more explosive and its products are less poor in silica. The volcanic activity on the floor of geosynclines is associated with the intrusion of peridotites and other rocks poor in silica, whereas the eruptions on the axes of geanticlines are associated with the intrusion of granites and related magmas rich in silica.

Groups of volcanic rocks, which were formed in a given area during certain periods of the geological evolution, show a definite chemical and mineralogical consanguinity. In that case we speak of consanguineous suites of igneous rocks. The members of these groups may have crystallized in the depth of the earth, such as the plutonites and the sub-volcanic rocks; or they may have solidified after their eruption at the surface.

The areas within which such suites of related igneous rocks occur, are called comagmatic regions of petrographic provinces.

The comagmatic regions are limited both in space and in time. This may be illustrated by the study of the Sunda Mountain System. The following structural elements and associated petrographic provinces can be distinguished in a cross section of this mountain system from Christmas Island in the Indian Ocean to the Karimundjawa islands in the Java Sea (Figures 10 and 11):

(1) The floor of the Indian Ocean, which is approximately 5,000 metres deep, and on the edge of which the extinct volcano of Christmas Island is situated. This volcano was active in the earliest Tertiary times and produced a series of lavas, belonging to the Atlantic suite. These Atlantic rocks are comparatively rich in sodium and occur elsewhere on earth invariably outside the mountain systems proper.

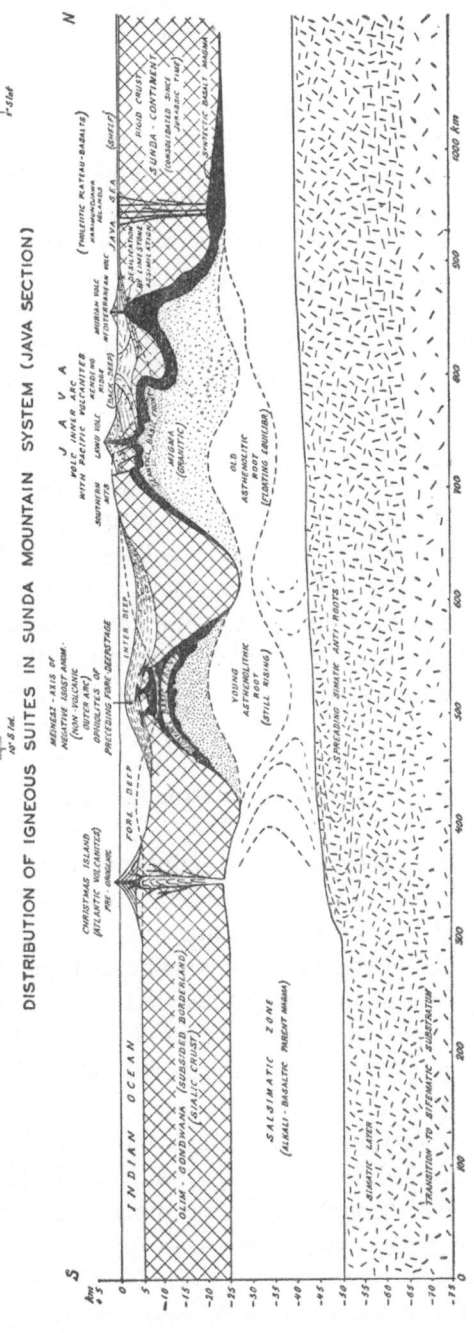

Fig. 10

Section across West Java and Christmas Island.

(2) Between Christmas Island and Java a deep occurs, which is divided into a northern and a southern basin by a submarine ridge. This ridge is a mountain chain in the process of formation. It is a geanticlinal ridge which is being pushed up from the deeper parts of a geosynclinal trough. This process is not yet complete, as is shown by the negative isostatic anomalies and the numerous earthquakes. We shall return to this matter in the following chapter. The submarine ridge south of Java is the extension of a similar ridge west of Sumatra, which has already given rise to a series of islands. Geological investigations in these islands have shown that in Late Mesozoic and Early Tertiary times ophiolitic igneous rocks were formed in this zone. At present this geanticline forms the non-volcanic outer arc of the Sunda Mountain System.

(3) The third petrographic province is formed by the geanticlinal ridge, which forms the axis of Sumatra and Java and supports a row of volcanoes. The products of these volcanoes belong to the Pacific suite. This latter group of volcanic rocks is a typical associate of mountain chains in their more mature phases of evolution.

The lavas of the Pacific suite are basalts, andesites, dacites and liparites. This type of volcanism is highly explosive, the volcanoes are consequently for the greater part built of magmatic material which has been scattered by the escaping gases and then deposited as tuffs etc. In this way the typical composite volcanoes originated, the cones of which are formed by alternating lava flows and beds of tuffs and breccias.

(4) Northern Java is occupied by a sedimentary basin situated between the geanticlinal ridge of southern Java and the central Sunda Land. In Quaternary times volcanic activity occurred in this zone, but the eruption products show transitions from the Pacific suite to varieties poorer in silica and richer in potassium, belonging to the Mediterranean suite of igneous rocks. The extinct volcanoes Muriah and Ringgit, situated near the north coast of Java, are composed of these less common types of lava.

(5) North of Java we find the old Sunda Land which is now for the greater part covered by a shallow sea. The Karimundjawa Islands are composed of strongly folded, presumably Lower Mesozoic sediments, intruded and covered by younger basalts. Similar basalts occur also as effusions in the eastern Lampong Districts near Sukadana, and at many other places in the border areas of the south-eastern

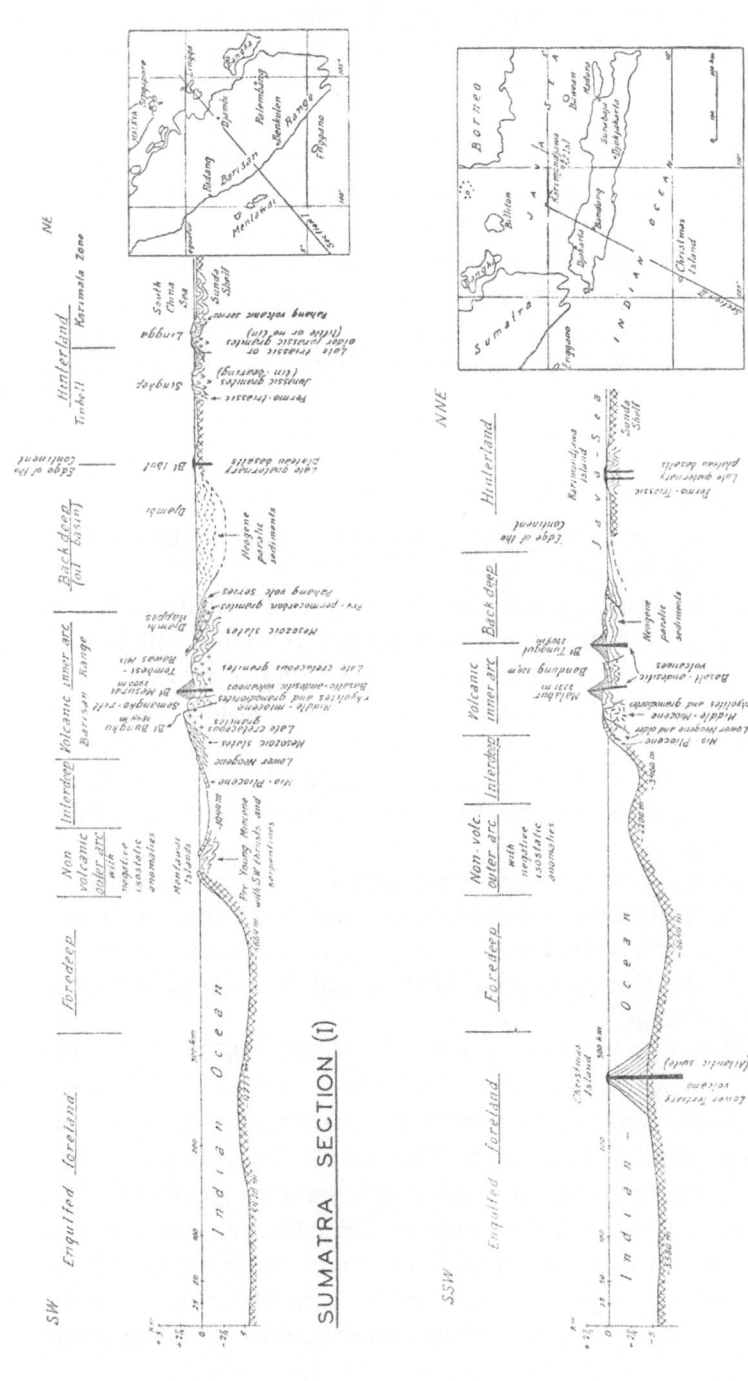

Fig. 11

Two schematical sections across the Sunda mountain system.

67

Asiatic continent. They are hardly different in composition from the basalts which occur in the Pacific suite. It is nevertheless striking that these basalts were not accompanied by varieties richer in silica (andesites-liparites), and that the eruptions were not as explosive as those of the Pacific volcanism. These young, often very voluminous, basalt effusions in the hinterland of the Sunda Mountain System belong to the group of the Plateau Basalts.

The following conclusions concerning the nature and distribution of the comagmatic regions within the Sunda Mountain System can therefore be drawn.

    I. In the "foreland" (in this case the Indian Ocean) the Atlantic suite occurs.

    II. In the non-volcanic outer arc ophiolites occur.

    III. The volcanic inner arc is characterized by the Pacific suite of volcanic rocks.

    IV. Varieties of the Mediterranean suite occur at the concave, inner side of the volcanic arc.

    V. The hinterland is characterized by effusions of Plateau Basalts.

The close relation between orogenesis and volcanism will be clear from the present distribution of these petrographic provinces. This relation becomes still clearer, if we do not only consider the present distribution of magmatic rocks, but also study the changes in distribution during the geological evolution. In other words, we must trace the distribution of the magmatic provinces both in space and in time.

In order to do this we must not only consider the products of external volcanism (the volcanites), but also study the distribution of the sub-volcanic and plutonic rocks. These have been exposed by erosion in the older mountain chains, and are accessible to our direct observation.

The effusions of sodium rich basalts (spilites) of the ophiolitic suite are mostly accompanied, in the depth of the earth, by intrusion of gabbros and silica-poor, often olivine-rich, plutonites such as peridotites, dunites, harzburgites, lherzolites and serpentines. The calc-alkali eruption products of the Pacific suite are accompanied by intrusions of diorites, quartz diorites, granodiorites and granites. These silica-rich varieties of plutonic rocks occur as batholiths, which are large, often dome-shaped masses of holocrystalline igneous rock, the boundaries of which cut rather arbitrarily through the formations into which they intruded. The intrusion, the so-called "mise en place", of these silica-rich batholiths took place during or shortly after the phases of active mountain building ,when mountain chains were pushed up from the geosynclinal tracts. This contrasts with the emplacement of silica-poor plutonites belonging to the ophiolitic suite, which were formed in the period when the crust was still subsiding. The plutonic rocks, therefore, also show a close relation between the phase of orogenesis on the one hand and their chemical composition and distribution on the other.

We shall now discuss two examples of the evolution of a geosynclinal trough into a mountain chain. The Meratus Mountains in south-eastern Borneo (see Table 3) began their life in Jurassic times, when the older crystalline crust began to subside. In the deeper parts of this geosyncline radiolarian ooze was deposited which is now found, together with the associated intrusive and extrusive ophiolites, as the Alino formation. Jurassic sediments here also occur in a neritic facies, the Paniungan formation. During Jurassic times the belt of the present Meratus Mountains was probably a deep sea trough. This trough was bounded on its eastern side by a rise in the earth's crust, situated where now Strait Macassar is. In this deep the Alino strata were deposited and more to the east, near what has been called the Strait Macassar Land, the Paniungan beds originated.

After the Alino and Paniungan strata had been folded and compressed, this trough with its contents of sediments and ophiolites was pushed up and formed a mountain chain in Early Cretaceous times. This was accompanied and followed by the intrusion of granites. External volcanism did, however, not yet occur.

This was followed in Cretaceous times by a new period of subsidence, during which the sea transgressed over the eroded mountain structures. In this sea Cretaceous formations were deposited (Orbitolina limestones, Manunggul formation). They are separated by an unconformity from the Jurassic sediments, the ophiolites and the Early Cretaceous granite intrusions.

This period of subsidence was terminated by a second uplift at the end of Cretaceous times. The resulting mountain chain was volcanic; lavas of the Pacific suite appeared at the surface. This volcanism continued into Eocene times. Intrusions of granite also took place. In Tertiary times a third phase of subsidence took place, and marine formations were deposited unconformably on the Mesozoic strata.

In Plio-Pleistocene times this unstable zone was subjected to a third uplift, which gave rise to the present-day Meratus Mountains which are no longer volcanic.

TABLE 3

Evolution of the Meratus Mountains in south-eastern Borneo

Orogenesis	Volcanism and plutonism
3rd impulse of uplift in Plio-Plei-stocene times	No external volcanism. The erosion has not yet sufficiently far progressed to establish the presence or absence of concomittant granite intrusions.
Quiet subsidence in Tertiary times	External volcanism persists into Eocene times.
2nd impulse of uplift at the end of Cretaceous times	External volcanism of the Pacific suite. The quartz dioritic fillings of vents are at present exposed.
Quiet subsidence in Middle and Late Cretaceous times	No external volcanism.
Ist impulse of uplift in Early Cretaceous times	No external volcanism. Granites intruded into the basic plutonites of the ophiolitic suite.
Subsidence in Jurassic times	Formation of ophiolitic intrusions and extrusions.

Summing up, we can say that in the Meratus zone geosynclinal subsidence began in Jurassic times, during which ophiolitic volcanites and plutonites were formed. This was followed (with intervals lasting some tens of millions of years) by three periods of uplift, each giving rise to a mountain chain which emerged from the geosynclinal zone; the first mountain chain was not yet volcanic, the second showed active volcanism and the third was no longer volcanic.

The first two uplifts, and by analogy presumably also the third (though of course this cannot be directly observed), were accompanied by the intrusion of granite into the crust.

A similar sequence of events is typical for all zones in Indonesia which were subjected to mountain building. The Barisan Mountains, for example, which form the backbone of Sumatra, show the following phases of development.

70

TABLE 4

Evolution of the Barisan Mountains in Sumatra

Orogenesis	Volcanism and plutonism
3rd impulse of uplift in Plio-Pleistocene times	Revival of the external basalto-andesitic volcanism. Intruding granites (3rd generation) caused violent eruptions of dacitic and rhyolitic pumice, e.g. those of Ranau and Toba.
Quiet subsidence in Mio-Pliocene times	External volcanism, mainly of basaltic and andesitic composition.
2nd impulse of uplift in Middle Miocene times	Violent eruptions of dacitic and rhyolitic pumice, especially along the Semangko fault zone on the top of the geanticline, are associated with the intrusion of granites (2nd generation).
Quiet subsidence in Oligo-Miocene times	Strong volcanic activity, mainly of basaltic and andesitic, but sometimes dacitic composition. Produced the "Old Andesite formation".
1st impulse of uplift in Late Cretaceous and Early Eocene times	No external volcanism. Intrusion of granites (1st generation)
Geosynclinal subsidence, especially in Late Mesozoic times (geosynclinal foredeep of a mountain chain N.E. of the Barisan zone).	Ophiolitic rocks intrude the geosynclinal sediments, e.g. in the Garba and the Gumai Mountains in southern Sumatra.

In Mesozoic times a subsiding tract was situated here, which, especially during Jurassic and Early Cretaceous times, was in the nature of a foredeep of a mountain system situated in the central Sunda Land. In this geosyncline marine sediments, but also ophiolites, were formed, which at present can be studied in the Garba and Gumai mountains in southern Sumatra.

The contents of this geosyncline were folded and compressed in Middle Cretaceous times. This gave rise to a formation, comprising several stratigraphic stages, which is known under the collective name of "Old Slates". Towards the end of Cretaceous times and the beginning of Tertiary times, this trough, with its contents was for the first time pushed up. In this way a mountain or island chain was vormed, called Primeval or Proto-Barisan. It did not show external folcanism, but masses of granite intruded the core of this ridge.

In Tertiary (Oligocene) times the Primeval Barisan subsided and the sea regained the greater part of its lost territory. Volcanic activity began in Oligo-Miocene times, especially in the southern part, where the so-called "Old Andesite formation" originated. In Middle Miocene times a second uplift took place ,which was accompanied by strong volcanic activity. At the same time granitic magma rose in the core of this mountain chain. The granites even penetrated into the Old Andesite volcanoes, and consequently must have approached the surface rather closely.

This Middle Miocene chain subsided again and the Pliocene ocean transgressed over large areas, as for example in Benkulen. Finally, a third period of uplift followed in Plio-Pleistocene times, which gave the Barisan range its present appearance. Violent eruptions of liparitic pumice, e.g. those of the Toba region, prove that the third uplift of the Barisan zone was also accompanied by the intrusion of granites.

In a similar way as the Meratus chain, the Barisan range passes through a first phase of geosynclinal subsidence accompanied by ophiolitic volcanism; this was followed by three periods of uplift, separated by periods of decreasing relief and subsidence. The difference between the two is that in the Barisan zone the orogenic phases succeeded each other faster than in the Meratus zone, the time intervals being 20–30 millions and 40–60 millions of years respectively. Another difference is the revival of the volcanism in the Barisan zone during and after the third period of uplift, whereas the third uplift of the Meratus zone created a non-volcanic range. These differences are caused by the different position of these zones in the structure of Indonesia. The Meratus zone is surrounded on all sides by other mountain systems, which hamper its development. The Barisan zone, however, is situated next to the basin of the Indian Ocean, which gives it much more freedom of development.

We have seen that in the section across the Sunda Mountain System certain comagmatic provinces are associated with certain zones of orogenesis.

We have also seen from the examples of the Meratus Mountains in south-eastern Borneo and the Barisan Mountains in Sumatra, that the successive phases of development of an orogenic zone are accompanied by certain suites of magmatic rocks. This leads to the principle of the succession in time of comagmatic suites belonging to a given structural zone.

The ophiolitic suite occurs during the geosynclinal or foredeep phase, and the igneous rocks of the Pacific suite accompany the periods of uplift, which transform the geosyncline into a mountain chain.

These two groups of internal and external volcanism are the principal companions of the orogenesis in Indonesia. The Atlantic suite and the Plateau Basalts occur outside the zone of active vertical movement; the former before the actual orogenesis has started, and the latter after this process has been completed. The Mediterranean suite comprises less common types of volcanic rocks, which originate only locally in the later phases of the orogenic development.

We shall now endeavour to trace the relations between orogenesis and magmatic phenomena, both in space and in time. To this end we must not only investigate the development of a certain structural zone, but also the adjacent zones.

We shall first study the changes in the distribution of certain comagmatic provinces in the course of time, and in the first place the distribution of the granites, which are the most characteristic plutonic rocks formed during orogenesis.

When we discussed the Meratus and Barisan ranges, we saw that each phase of uplift was associated with the intrusion of granitic rocks into the core of the geanticline, even though not all periods of uplift were accompanied by external volcanism.

If we now study the age of the granitic intrusions in the Archipelago, it appears that zones of granites of different ages can be distinguished (see Fig. 12).

In the Sunda region the oldest granites occur in a zone which stretches from the Anambas Islands to the Schwaner Mountains in central Borneo. Their age is post-Permo-Carboniferous and pre-Triassic. On both sides of this zone, namely in the Natuna zone to the northeast and in the Karimata zone to the south-west, zones are found where the age of the granites is post-Triassic and pre-Late Jurassic. More to the south-west, a zone of granites which have produced tin ores, occurs. It runs across the western part of the Malayan peninsula and the islands Singkep, Bangka and Billiton (see Fig. 13). The granites in this Tin zone are presumably still younger, namely Late Jurassic.

These Permo-Triassic and Jurassic granites form, as it were, the

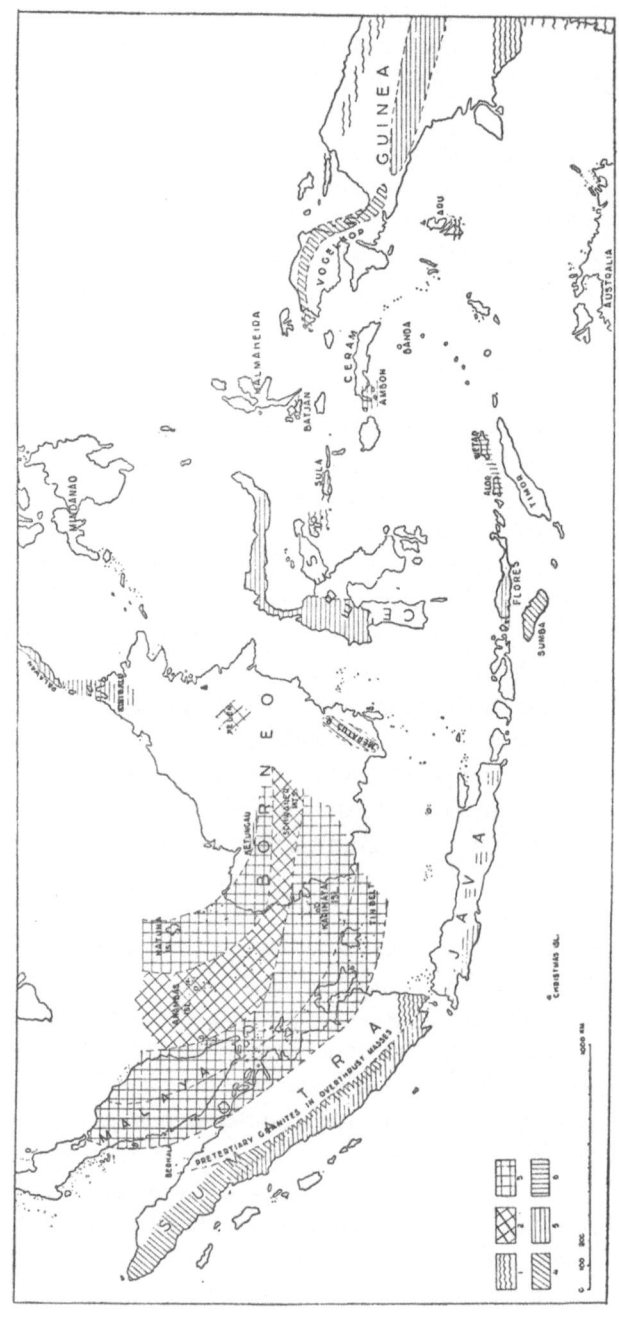

Fig. 12

Outline map of the East Indies, showing the distribution of granitic rocks in orogenic belts.
1. Crystalline basement complex granites. 2. Permo-triassic granites. 3. Jurassic granites. 4. Cretaceous granites. 5. Mid tertiary granites. 6. Late Tertiary granites.

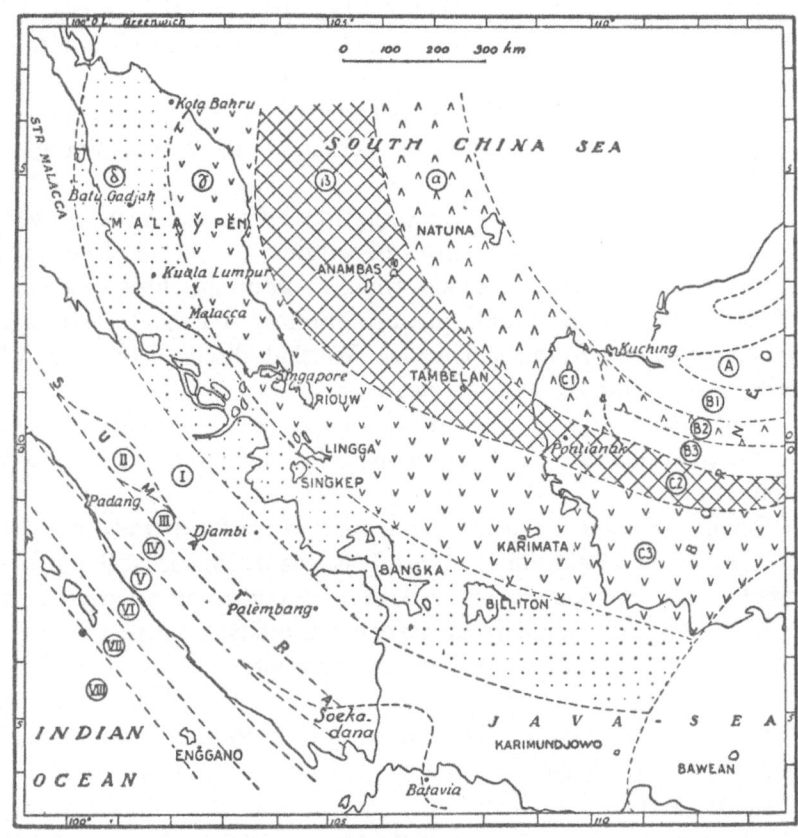

Fig. 13

Structural belts in the Sunda area.

Old mountain systems

In the western part
(Sunda shelf area, Malay Peninsula)
α = Natuna zone
β = Anambas zone
γ = Karimata zone
δ = Tin zone

In the eastern part
(Western and Central Borneo)
= C_1 (Chinese districts)
= C_2 (Schwaner Mountains)
= C_3 (Ketapang-Matan district)
= Continuation in the Java Sea.

Young mountain systems

The Sunda mountain system in the southwestern part.
        I = Backdeep (oil basin of East Sumatra)
II + III = Nappes on the eastern flank of the Barisan ⎫
      IV = Barisan with Cretaceous granites ⎬ (Volcanic inner arc)
       V = Barisan with Miocene granites ⎭
      VI = Interdeep
     VII = Non-volcanic outer arc
    VIII = Foredeep

75

frame of the present Sunda Land. The intrusion of these granites made the central Sunda region rigid, and transformed it into a continental block. Later, in Cretaceous and Cainozoic times, the active orogenesis moved farther outwards, to north and north-west Borneo and to the Java-Sumatra section of the Sunda Mountain System.

East of the Barisan range in Sumatra we find a number of over-thrust masses, in which granites of various ages occur. One finds granites of pre-Carboniferous age, which make up the oldest basement of the central Sunda Land, but also granites of Late Jurassic age which have come from the Tin zone. We shall discuss in chapter VI how these granites came to be there.

In the Barisan range itself we find exposures of granites of Late Cretaceous and Middle Tertiary age. The latter are also known from Java.

Finally we reach the geanticlinal zone which forms the outer arc of the Sunda Mountain System. This geanticline is still for the greater part submerged; only west of Sumatra it appears as a number of islands where its nature can be studied. Granite intrusions are lacking. This zone, however, is characterized by a striking deficiency of gravity, a fact which we shall discuss in the following chapter on geophysics. This indicates that rocks or magmas of relatively small density are present in and below the crust in this zone. It is therefore likely, that also in the core of this young geanticline granites are intruding. The development has not yet progressed so far that granite batholiths are exposed by erosion and are accessible to direct observation.

It appears, therefore, from the distribution of the granites in the Sunda region, that the intrusions started in a central zone, and in younger periods of the earth's history they shifted in a lateral direction.

A similar result is shown by the distribution of the ophiolitic suite.

Of the older ophiolites in the central Sunda region, the distribution is only imperfectly known. This is caused by the later events, which often strongly metamorphosed and transported these volcanites, and also because reliable determinations of age are scarce. The Devonian marine sediments of the Telen region are accompanied by diabases (i.e., basalts, partly changed in their chemical and mineralogical character), and the Permo-Carboniferous sediments of the Danau formation in western Borneo and the Natuna Islands are also accompanied by basic volcanites (the Pulu Melaju Basalts). These eruptive rocks, which are basic and accompanied by marine sediments, may be considered as ophiolites. The volcanic products which accompany the Permo-Carboniferous sediments in the Malayan peninsula are not very basic; the Pahang Volcanic Series has rather an intermediate to acid composition and resembles the Pacific rocks. In the western Riouw Archipelago strongly folded lustrous slates and amphibolites occur. This formation is separated by an unconformity from the overlying Riouw formation, comprising the Permian and Triassic. It is not impossible that these amphibolites belong to a pre-Permian ophiolitic suite.

Nor are Early-Mesozoic ophiolites clearly developed in the Sunda region. It is possible that the basalts and diabases which occur as intrusions in the Riouw formation, but which are older than the Jurassic Tin granites, represent Early-Mesozoic ophiolitic volcanism.

Late-Mesozoic ophiolites definitely do not occur in the central Sunda region, but they are known from the border zone in Sumatra and Java where they are associated with formations of Early Cretaceous age.

Tertiary ophiolitic rocks are known from the outer arc, as serpentinized intrusions in Oligocene formations of the islands Nias and Sipura.

The ophiolites therefore become increasingly younger when we move from the central parts of the Sunda Land in a south-westerly direction.

A similar law is exhibited by the external volcanism, which produces the Pacific suite of eruptive rocks. The information on their distribution is much less complete than on that of the granites. This is partly due to the fact that the magma which took part in the orogenesis, could not always reach the surface, and secondly because the eruption products were later on often removed by erosion. It is therefore not always possible to decide whether or not old mountain ranges have passed through a phase of external volcanism.

Volcanic formations of Permo-Triassic age are known in the Natuna and the Karimata zones. In Early Cretaceous times there was volcanic activity of a Pacific character in eastern Sumatra (Saling facies of the Lower Cretaceous). In Tertiary and Quaternary

TABLE 5

Scheme, showing the distribution of the comagmatic suites of igneous rocks in the Sunda region.

	Indian Ocean (foredeep)	Islands west of Sumatra (outer arc)	Western Sumatra (Barisan)	Eastern Sumatra (Oil basin)	Tin Zone	Karimata Zone	Anambas Zone
Devonian							Ophiolitic volcanism
Permo-Carboniferous			.			Ophiolitic volcanism / 1st Granite intrusion?	1st Granite intrusion? Pacific volcanism? / 2nd Granite intrusion
Triassic					Ophiolitic volcanism	Pacific volcanism / 2nd Granite intrusion	
Jurassic					Granite intrusion		
Cretaceous			Ophiolitic volcanism / 1st Granite intrusion	Pacific volcanism / Granite intrusion			
Tertiary		Ophiolitic volcanism	1st Pacific volcanism / 2nd Granite intrusion / IInd Pacific volcanism / 3rd Granite intrusion				
Quaternary	Ophiolitic volcanism?	Granite intrusion?	IIIrd Pacific volcanism	Plateau-basalts			

times the volcanic activity migrated to the Barisan range in Sumatra and the geanticline of Java. Finally, it may be predicted that the outer arc of the Sunda Mountain System will become volcanic in the course of time, i.e. in the geological future.

The external volcanism, the products of which belong to the Pacific suite, therefore also becomes younger when we move away from the central part of the Sunda Land in a south-westerly direction.

The relation between the structural belts and the igneous suites in the Sunda region, as shown in table 5, conforms to the more theoretical scheme of table 5a in which the stages of orogenic and petrogenic evolution are correlated (from the author's paper 1950 a, p. 211, and 1953).

TABLE 5a

Theoretical scheme of the relations between petrogenic and orogenic stages of evolution

Petrographic provinces	Orogenic Zones	Zonal stages of orogenic evolution	Stages of evolution of the orogenic system				
			Prefatory	Embryonic	Young	Early mature	Mature
(I) Atlantic suite	Foreland (1)	Pre-orogenic	X	X	X	X	X
(II) Ophiolitic suite	Foredeep (2)	Geosynclinal		X	X	X	X
(III) Pacific suite	Geanticline (3)	Orogenic (sensu stricto)			X	X	X
(IV) Mediterranean suite	Backdeep (4)	Late orogenic				X	X
(V) Tholeiitic plateau-basalts	Hinterland (5)	Post-orogenic					X

Relations between volcanism and orogenesis.

The foregoing analysis of the distribution of the volcanic phenomena in space and time, shows that the Sunda region possesses a pronounced zonal structure. The central zone (Anambas Islands to Schwaner Mountains) is the oldest. From the central zone, internal and external volcanism migrated gradually in a lateral direction. This process can be followed at its clearest and completest in a south-westerly direction, as is shown by Table 5.

In the eastern part of the Archipelago a similar close relation exists between volcanism in its wider sense, and orogenesis.

The necessity of a close relation between volcanism and orogenesis will be clear when we realize what the significance of these processes is for the development of our planet.

When the earth was born, over three thousand millions of years ago, it possessed an enormous quantity of energy. This energy occurs in various forms, such as heat, chemical forces between the elements which constitute the earth, nuclear energy (natural radio-activity), etc. These stores of energy strive after an internal equilibrium.

This internal equilibrium has not yet been reached by the earth, because it is disturbed by its cooling. This cooling, however, is not quite so simple as the decrease in temperature of a kettle of warm water or other hot objects in our daily environment.

In the first place, it is not necessarily true that the process is uninterrupted and of the same intensity everywhere at the surface. In the second place, every change in temperature in the earth causes changes in the chemical equilibrium. The chemical processes which occur because of this cooling, are exotherm, i.e., they produce heat. They tend to counteract the effects of the cooling of the earth.

This tendency towards an internal equilibrium is associated with the migration of matter and the energy it carries. These physico-chemical chain reactions must be considered as the source of the internal energy of the earth, which expresses itself at the surface in volcanism and tectonics.

Volcanism in its wider sense is the chemical aspect of the transfer of energy from the depth to the surface, and tectonics is its mechanical aspect. Both have the same fundamental cause, i.e. the transfer of energy from the depths of the earth to the surface.

Volcanism in its wider sense could also be defined so as to comprise all phenomena relating to the migration of matter and the physico-chemical energy it carries. This migration can occur in concentrated form, as magma, or in a more dispersed form, as emanation. If it appears on the surface we call it lava and exhalation (of volcanic gases) respectively. The chemical aspect of the crustal evolution is not always as clear to man as is the mechanical aspect. If the earth's crust is thick, the chemical processes which occur in the depth, and which are taken to belong to volcanism in its wider sense, cannot easily be observed at the surface. The result is that the mechanical expression of the transfer of energy (the tectonic deformations) in this case predominates. In other places, where the crust is thin, volcanic processes often predominate, as for example in the island arcs along the western and south-western side of the Pacific Ocean.

In Indonesia the conditions are such that both chemical and mechanical aspects of the geological evolution can be clearly observed. This makes the geology of this region of more than local interest.

Transitions between volcanic and tectonic relief
We shall now illustrate the close relation between volcanism and tectonics by a short discussion of the relief forms which are caused by volcanic eruptions, and are connected to the tectonic forms of the relief by intermediate stages.

The relief created by external volcanism is the result of constructive and destructive factors. The balance between the supply of volcanic material from the depth on one hand, and the removal from the centre of eruption on the other, determines the resulting form. If the supply exceeds the withdrawal, i.e., if the balance is positive, positive volcanic structures, such as cones, etc., originate. If the supply is less than the withdrawal, i.e. the balance is negative, negative volcanic forms originate (depressions). Within both principal classes, (A) the positive and (B) the negative forms, we can distinguish purely volcanic and volcano-tectonic types.

(A) The normal volcanic structures originate when the amount of material coming from the depth equals that produced at the centre of eruption; the volcanic products are partly or entirely piled up around the centre of eruption. This is how the well known volcanic cones originate.

If, however, the supply of magma from the depth exceeds the production of the volcanic vent, the surface is pushed up. The occurrence of this phenomenon, which takes place before the actual eruption, has been demonstrated by precision-levels. The same phenomenon, on a somewhat larger scale, is illustrated by the Merbabu volcano in Middle Java. The vent of this volcano became blocked up, the cone was pushed up by the advancing magma and finally radial faults originated, through which the lava appeared at the surface (Fig. 14).

The lignite bearing strata of the Middle Palembang formation, have been pushed up in a similar manner by andesitic lavas forming

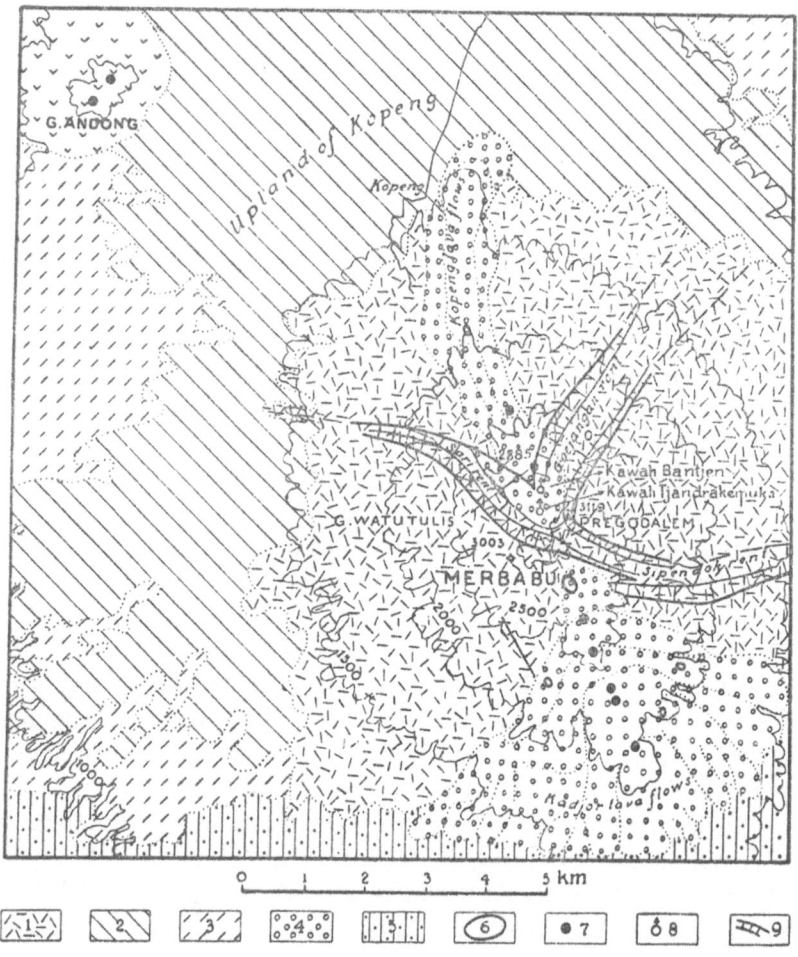

Fig. 14

Geological sketchmap of Mt. Merbabu.

1. Merbabu cone, mainly basalto-andesitic lavas and breccias. 2. Highland of
Kopeng, covered by volcanic ashes. 3. Foot of the Merbabu cone, chiefly lahar
breccias and subordinately intercalated lava flows. 4. Youngest lava flows of Mt.
Merbabu with the centres of eruption, arranged along a NNW-SSE line, forming
part of the major transverse zone of volcanic activity from Ungaran to Merapi.
The Kopeng lava descended northward and the Kadjor lavas flowed southward.
5. Northern foot of the Merapi volcano, covered with Merapi ashes. Selo pass
between Mt. Merbabu and Mt. Merapi. 6 and 7. Centres of eruption of the Kopeng
and Kadjor lava flows, and the small Andong volcano, NW of Mt. Merbabu. 8. Mofettes and very weak exhalations of sulphuretted hydrogen on the summit (Kawah
Bantjen and Kawah Tjondrokemuko). 9. Volcano-tectonic rents or sector graben;
presumably caused by a final doming up of the Merbabu cone, when the central
vent was definitely clogged by a lava plug and the magma, pressing upward, could
not reach the surface anymore.

82

Physiography of North Sumatra, showing the Batak
culmination of the Barisan range and the longitudinal
Semangko graben along the crest of this range.

a. Butung Sumpur valley
b. Batang Gadis – Batang Ankola valley
c. Batang Toru valley
d. Toba cauldron
e. Alas valley
f. Blang Kedjeren basin
g. Atjeh valley

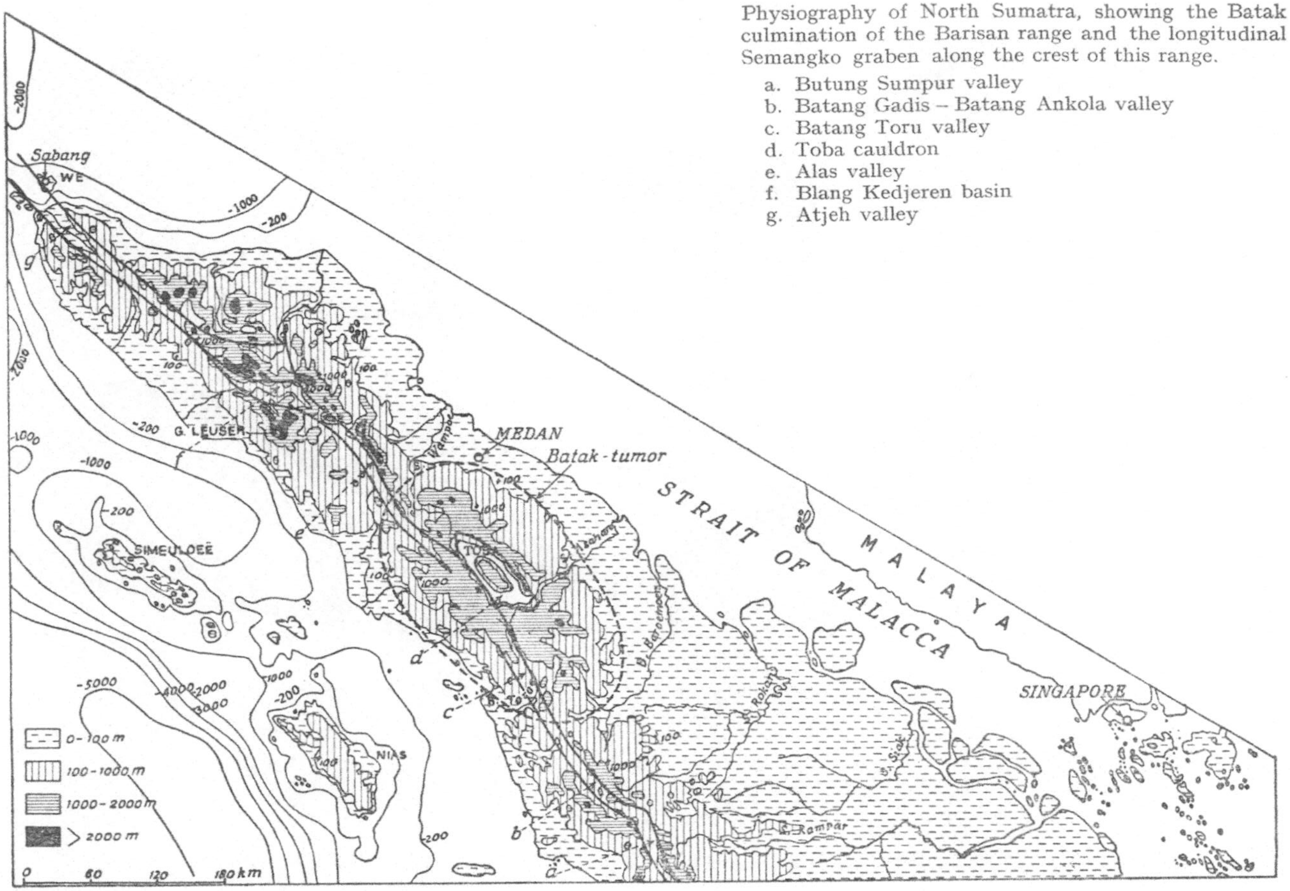

Fig. 15

83

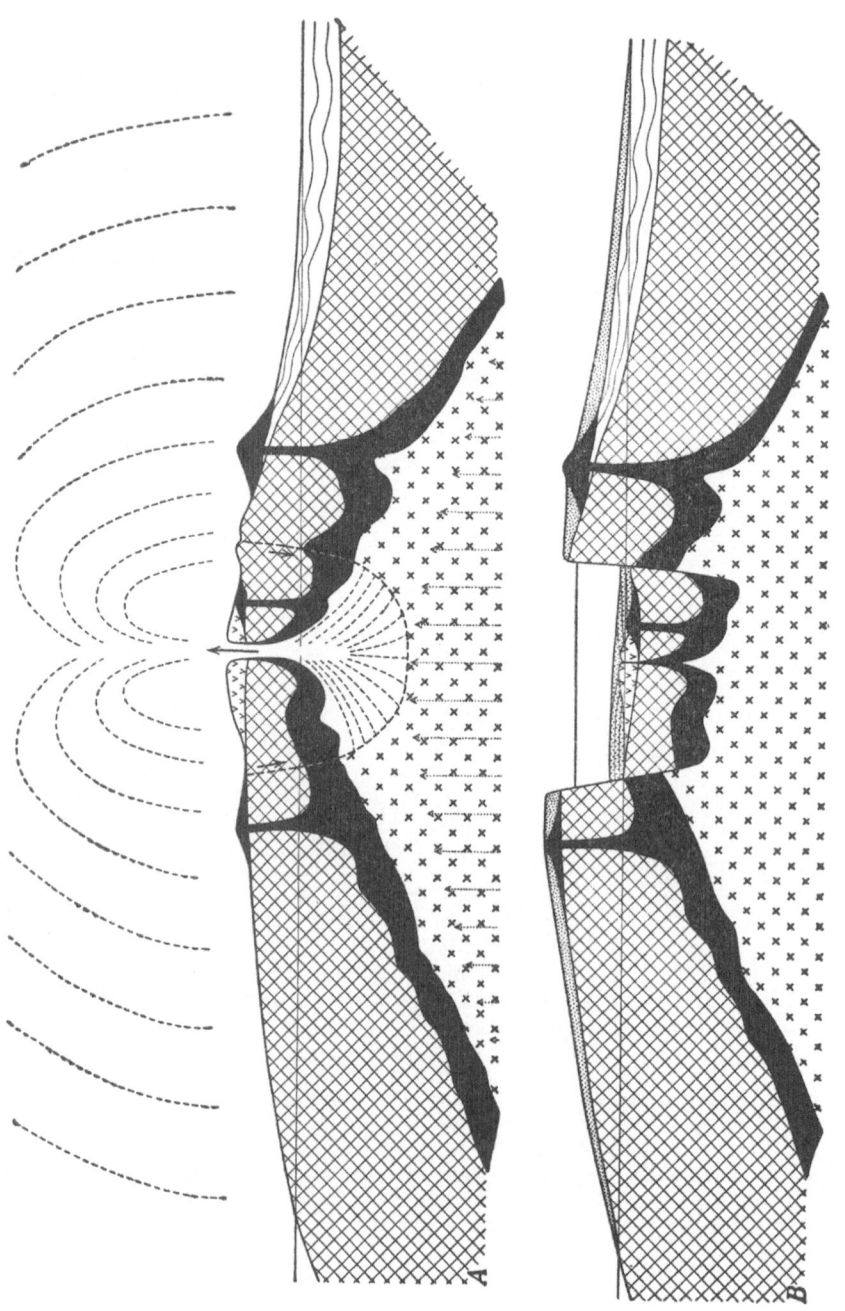

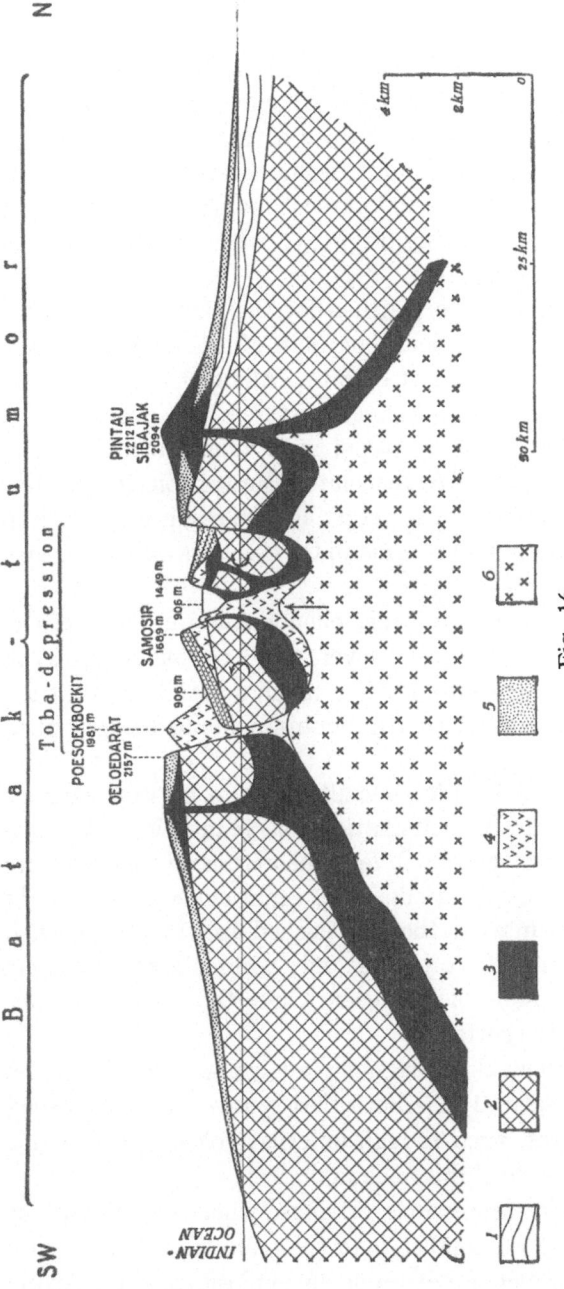

Fig. 16

Three schematic sections across the Batak tumor, showing the consecutive stages of formation of the Toba cauldron.

1. Neogene marine deposits of the coast of Medan.
2. Pre-tertiary basement complex.
3. Basaltic and andesitic magmas and eruption products.
4. Welded rhyolite tuffs and breccias of Samosir and the Prapat-Porsea Peninsula (Ignimbrites). The Pusuk Bukit volcano, which is younger, consists of hypersthene andesite.
5. Loose rhyolitic-dacitic Toba tuffs.
6. Granite batholith of Toba.

16a. Paroxysmal blowing out of the top part of the migmatic granite batholith, which formed the core of the Toba tumor and which pushed up the overlying geanticlinal vault. This doming up was accompanied by the formation of tension fissures in the roof, along which the gas-laden granitic magma could escape.

16b. Collapse of the vault and formation of the Toba cauldron.

16c. Tilting of the Samosir and Prapat-Porsea cope stones by subsequent uptrusions and continued basalto-andesitic activity (Sibajak, Pusuk Bukit).

laccoliths. The volcanic complex of Bukit Mapas, south of Baturadja in southern Sumatra, is a horst, pushed up by the magma.

Geanticlinal volcanic ridges sometimes show typical culminations, called tumors, as for example the Gedongsurian tumor in southern Sumatra and the Batak tumor in northern Sumatra (Fig. 15, 16 and 17). Granitic magma reached the surface as incandescent flows of tuff, or as eruptions of pumice and tuff, through tension faults near the tops of these tumors. These tumors reach more than one thousand metres altitude, and presumably originated when the earth's crust was pushed up by rising magma.

This assumption seems also justified for the Barisan range as a whole. It appears, that every period of uplift of this mountain chain was accompanied by intrusion of granites. Along the entire length (1650 kilometres) of the geanticline, a system of tension faults developed, called the Semangko zone, with which intrusions and extrusions of granitic magma are associated. Many geologists assume that such large folds of the earth's crust are the result of tangential compression. The intrusion of magma in their view is only a passive, subordinate phenomenon.

The relation between the volcanic phenomena and the tectonic evolution of the Barisan range, however, raises serious doubt as to the correctness of this assumption. It is much more likely, that the rising magma is responsible for such mountain chains being pushed up.

The volcanoes of the Pacific, calc-alkali, suite mostly occur as parasitic structures on the culminations of geanticlines. In such cases, batholithic intrusions, which feed the external volcanism, are present in the core of the geanticline. The geanticlines which do not show external volcanism when they are pushed up, also possess such plutonic intrusions in their cores. The only difference with volcanic geanticlines is that in non-volcanic geanticlines the magma crystallized before reaching the surface.

If we accept the view that magma plays an active rôle in the orogenesis, it is also natural to believe that geanticlines are pushed up by magmatic forces, even though external volcanism does not always prove the presence of this magma.

(B) Purely volcanic depressions originate when the supply of magma from the depths is smaller than the removal of material by eruptions. This can happen, for example, by explosions. The volcanic

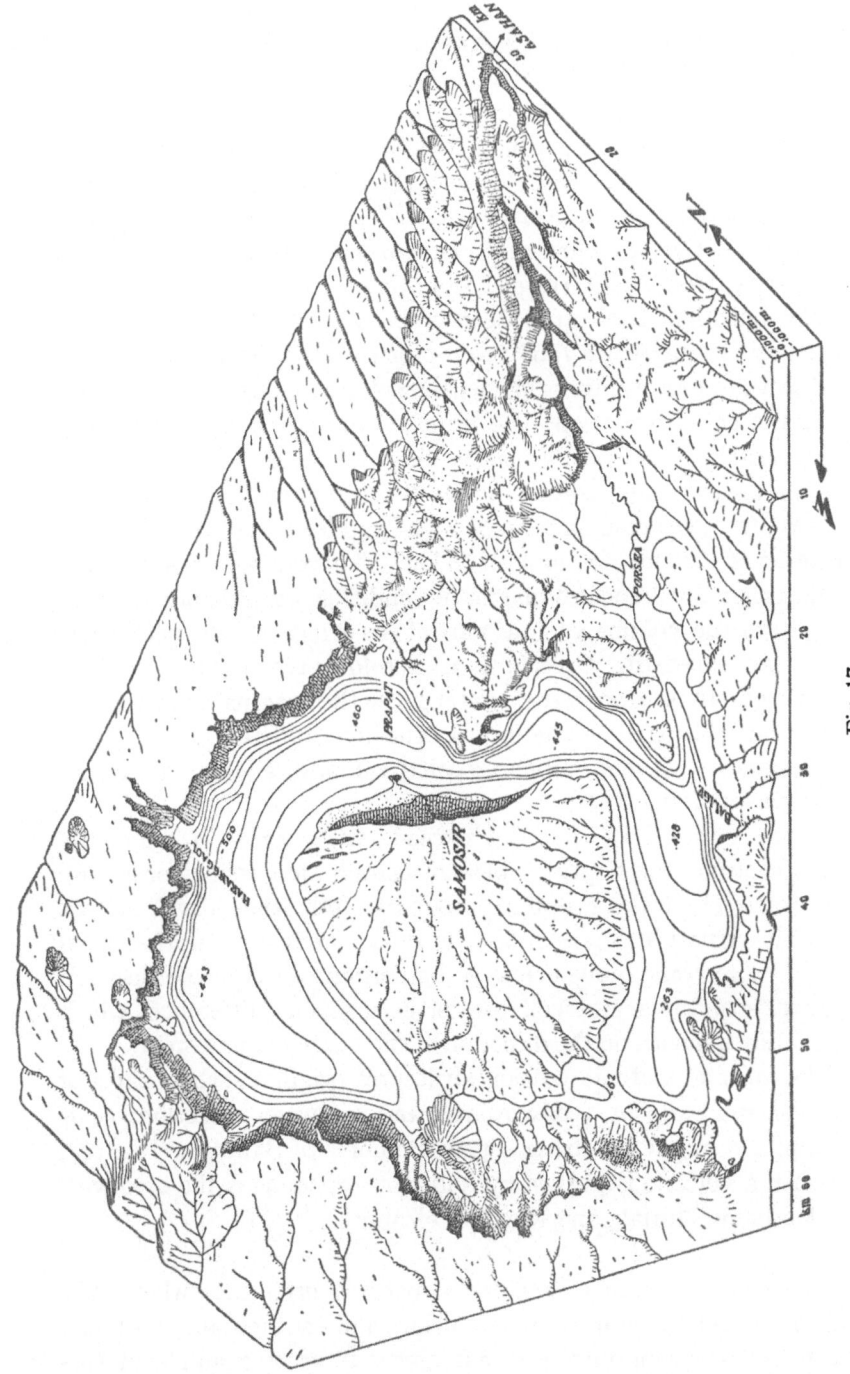

Fig 17.

Isometric block diagram of the Toba cauldron. (Depth figures of the Toba lake in metres).

material is broken up and scattered, and craters originate. It is, however, also possible that the eruption is followed by a collapse of the volcanic structure. In that case we speak of calderas. Calderas mostly have a much larger diameter than the ordinary craters. The subterranean room for the collapse of the volcano can originate when the magma is blown out of the magma chamber by violently escaping gases (Krakatau), or when the lava is drained away at a lower level (Zandzee caldera in the Tengger Mountains).

The question has often been discussed, whether during the Krakatau eruption of 1883 the top of the volcano was broken up and scattered, or whether the volcano collapsed, during and directly after the actual eruption. The principal argument against the top being destroyed, is the observation that the material of the eruption, for the greater part, consists of fresh ash, which must have come directly from the magma chamber. The pumice tuff of the Krakatau eruption of 1883 has a volume of 18 km^3. This volume, when reduced to the original volume in a magmatic condition, corresponds approximately to the volume of that part of the group of volcanoes which had disappeared after the eruption. Therefore presumably the following events took place: the content of the magma chamber was blown out through the channel of eruption, and the group of volcanoes, now without support, collapsed and a caldera-like depression was formed.

If the removal of material exceeds the supply of magmatic material from the depths, volcano-tectonic depressions originate. The removal of the eruption products and other parts of the volcanic structure, in this case, is done partly by eruptions, and partly by tectonic movements. A distinction can be made between tension structures and collapse structures, a difference which ultimately depends on the direction of the tectonic movements.

The volcano-tectonic tension structures originate when parts of the volcano slide in a downwards and sideways direction, as is exemplified by the Lembang depression north of Bandung (Fig. 18); or when a volcano is pushed up and burst open, as is the case with the Merbabu, Muriah, and Ringgit volcanoes.

The volcano-tectonic collapse structures originate, when only part of the magma appears at the surface and another part is drained away by injections into adjacent areas. A good example of this

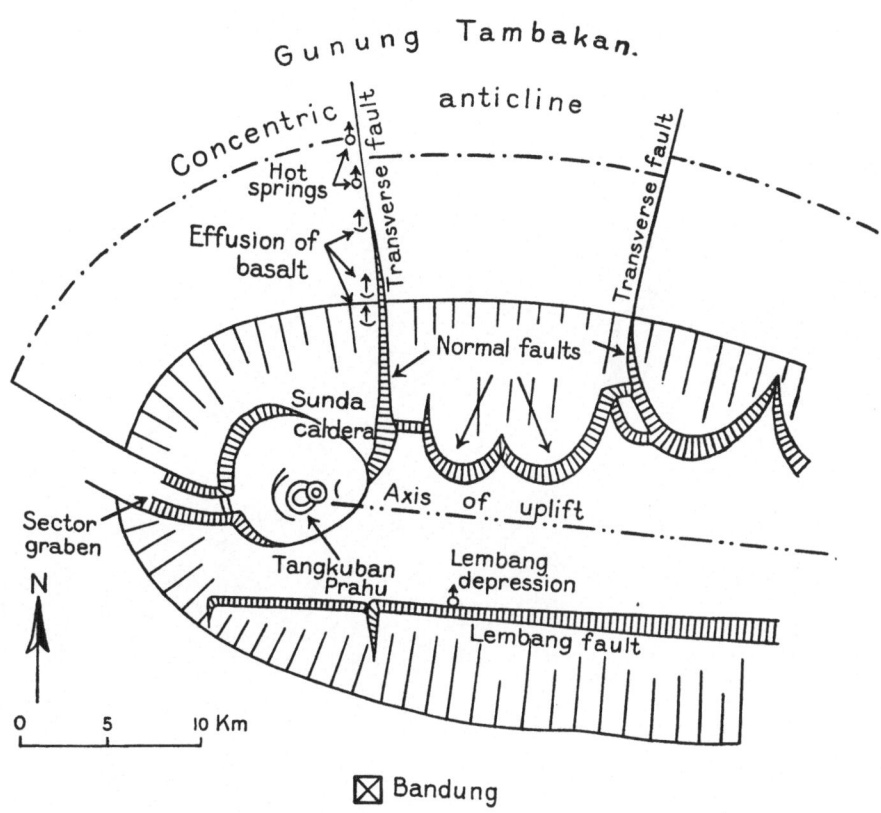

Gunung Tambakan.

Concentric   anticline

Hot springs

Effusion of basalt

Transverse fault

Transverse fault

Normal faults

Sunda caldera

Sector graph

Axis   of   uplift

Tangkuban Prahu

Lembang depression

Lembang fault

N

0   5   10 Km

⊠ Bandung

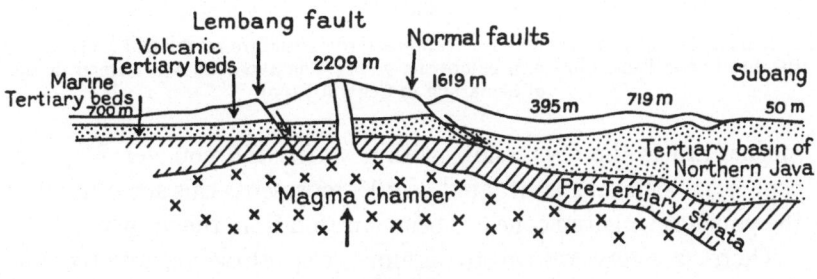

Lembang fault

Volcanic Tertiary beds

Marine Tertiary beds 700 m

2209 m

Normal faults

1619 m

Subang

395 m   719 m   50 m

Tertiary basin of Northern Java

Magma chamber

Pre-Tertiary strata

Fig. 18

Sketch and section to explain the structure of the volcanic complex North of Bandung (West Java).

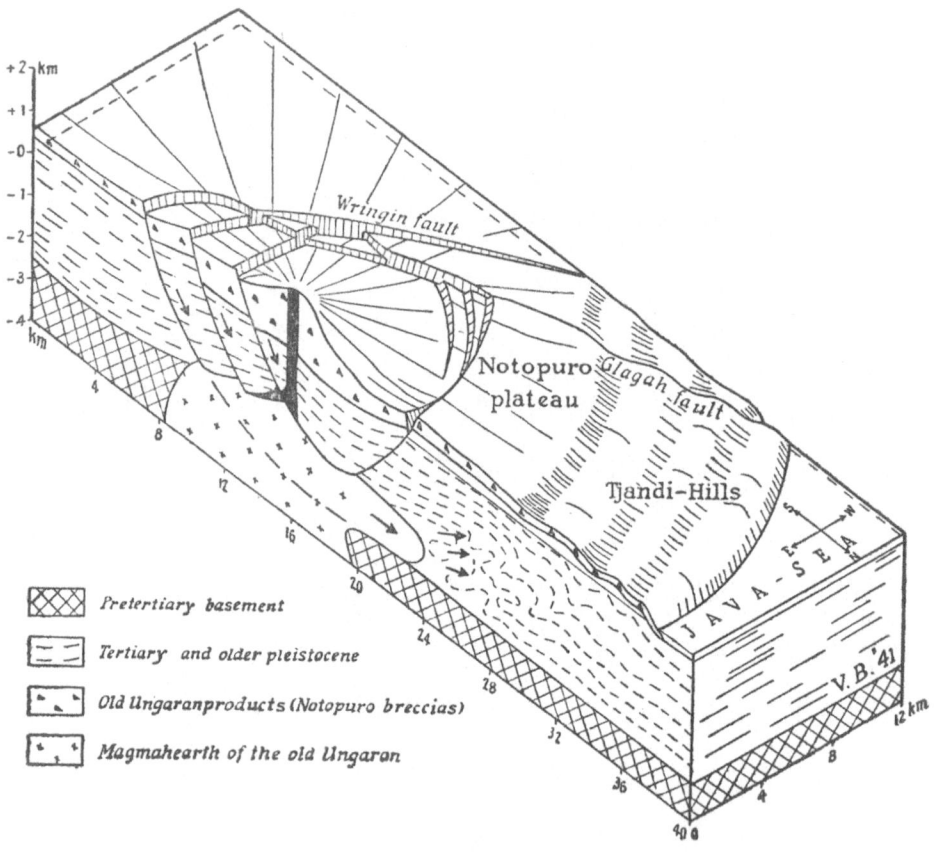

Fig. 19

Schematic block diagram of the volcano-tectonic structure of the Old Ungaran at the end of the Pleistocene and before the young Ungaran cone was formed (South of Semarang in Central Java).

mechanism is the depression, within which the younger Ungaran cone near Semarang is situated. Fig.19 represents this structure, but the younger Ungaran cone has been omitted from the drawing.

There is every reason to assume, that these volcano-tectonic depressions are associated with, and grade into purely tectonic collapse structures when purely tectonic structures are formed. The space for the subsiding crustal block is only for a small part, or not at all, provided by external volcanism, but is the result of magmatic movements in and below the earth's crust. For instance, the geanti-

90

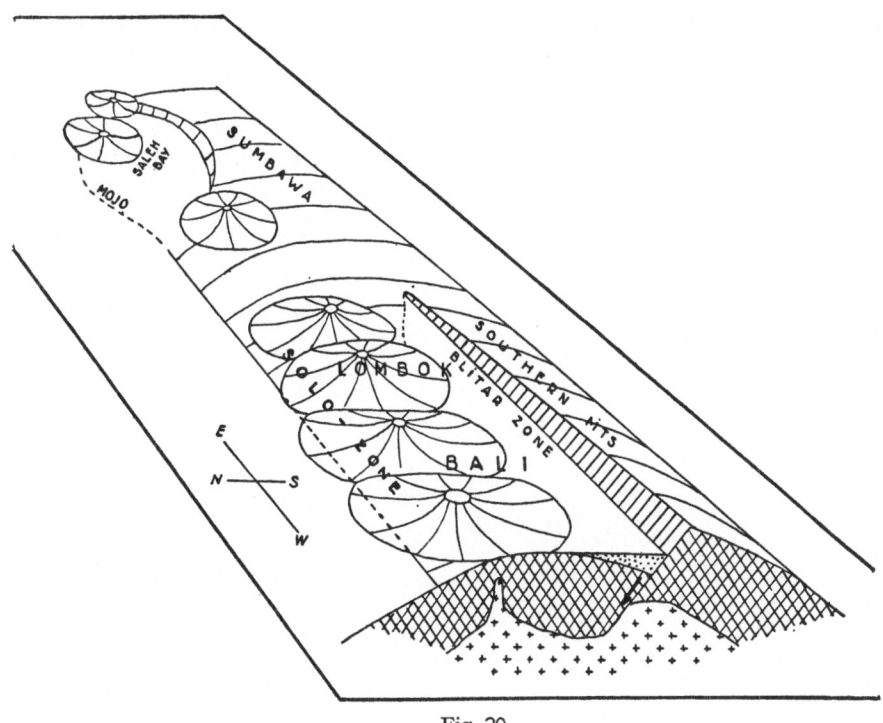

Fig. 20
Perspective sketch of the western part of the Lesser Sunda Islands.

cline of Java, Bali and Lombok shows subsidence structures which, for the greater part, are of purely tectonic origin. It appears that in Quaternary times the culmination and northern flank of the geanticline moved downwards and northwards relative to the southern flank, along a longitudinal series of faults. Many active volcanoes of this geanticline are situated on or near this fault zone (Fig. 20). The total volume of the material produced by these volcanoes is small compared with the volume of the depressions (graben) on the back of the geanticline.

This subsidence of the cope-stone blocks of the geanticlinal vault can, therefore, not be explained by external volcanism; it was volumetrically compensated for the greater part by displacements of magma in and below the crust, whilst volcanic eruptions were responsible only in a minor degree.

Another example is the collapse of the dome shaped tumor in the

Sunda Straits area. Here the subsidence of crustal blocks was accompanied by violent volcanic eruptions (Krakatau is only an after-effect), but the removal of material which caused this collapse and compensated it volumetrically, took place mainly in the depths. We shall discuss the origin of the Sunda Straits more fully in the next chapter as an example of gravitational tectogenesis.

IV. TECTONICS

What is tectonics?

In the previous chapter we pointed out, that volcanism and tectonics are both expressions of forces inside the earth. Both are associated with the transfer of energy to the surface, which energy is finally lost to the universe in the form of radiation. The volcanic activity represents the chemical aspect of this transfer of energy, whereas tectonics represents its mechanical aspect.

The tectonic deformation of the earth's crust — a process which has also been called diastrophism — is responsible for the formation of continents and oceans, plateaus and mountains, deeps and oceanic basins, as well as the folds and dislocations in the sedimentary strata. These processes can be divided into two principal groups, namely the primary, relief-creating movements, and the secondary, relief-destroying reactions, which tend to counteract the first.

All vertical movements of parts of the earth's crust which cause deviations from the ideal, spheroidal shape of the earth, belong to the primary tectogenesis.

These primary, radial movements of the earth's crust, are often subdivided into two groups, namely epeirogenesis and orogenesis.

By epeirogenesis we mean the relatively slow, rising and subsiding, movements of extensive regions of the earth's crust. The subsidence of the eastern part of the Archipelago (the Moluccas), for example, is an epeirogenic process which started in the youngest geological period. There are arguments in favour of the assumption that in the South China Sea, east of the Philippines, north of New Guinea and south of Sumatra, Java and the Lesser Sunda Islands, borderlands have existed, which now have sunk to oceanic depths and partly are still subsiding. The diameter of such areas amounts to more than 1,000 kilometres, and the rate of the epeirogenic emergence or subsidence is of the order of 0.1–1 mm per year.

By orogenesis is meant the process of mountain building. It cannot always be clearly distinguished from epeirogenesis, but the differential vertical movements happen in general much faster (approximately 0.1–1 cm per year). The diameter of the rising and sinking areas on the other hand, is much smaller, namely from one hundred to a few hundreds of kilometres.

The process of orogenesis, in Indonesia, is responsible for the major folds of the earth's crust: the geanticlines and geosynclines, and also for the horst mountains and the submarine troughs, which locally create great contrasts in altitude.

Primary tectogenesis is the effect of processes in the depth, which strive towards equilibrium and are accompanied by mass displacements. The primary deformations of the earth's crust are therefore closely associated with the plutonic processes.

The primary tectogenesis, in its turn, disturbs the equilibrium of the earth's crust near the surface. The topography above the geoid possesses too much potential energy and the basins have too little. This gives rise to stresses and the primary relief tends to decrease by gravitational flowage. When sediments and parts of the crust, in uplifted areas, slide sidewards and are compressed, folded and overthrust in the areas of subsidence, we speak of secondary tectogenesis or gravitational tectogenesis.

What is gravitational tectogenesis?
In this chapter we shall discuss some examples of gravitational tectogenesis in Indonesia; but before doing so, we shall first discuss the nature of the secondary tectogenesis and the types which can be distinguished (see Fig. 21).

After primary tectogenesis has created relief, this can be lessened or destroyed in two different ways. In the first place, erosion may attack the higher parts and the rock is transported to the adjacent lower areas, partly in solution, partly in a finely divided form. The rock is therefore dispersed, and transported along the surface from higher to lower areas. Theoretically it is possible that this exogene denudation is exclusively responsible for the transformation of mountain chains into peneplains. But the transport of material also occurs in a more concentrated form. The mountains collapse, like idols on feet of clay, and break up while their highest parts move downwards. It may also happen that raised series of plastic sediments slip like a cloak from the shoulders of the rising basement. In that case we speak of tectonic denudation.

The stresses, which the topography sets up in the crust, are not restricted to the superficial strata. The trajectories of maximum shear stress pass also through the deeper parts of the high areas, and come again to the surface in the adjacent lower regions. The destruction of a mountain is therefore not necessarily restricted to such superficial processes as erosion, creep, landslides and the gliding of sedimentary series. It depends on the physical properties of the material if, and where, the stresses will lead to deformations, i.e. to secondary tectogenesis.

The distinction which has just been made between primary tectogenesis which creates relief, and secondary tectogenesis which destroys it, leads to a bicausal interpretation of the mechanism of earth movements. This is fundamentally different from the unicausal interpretation of tectonics, which explains all earth movements by a hypothetical tangential pressure in the earth's crust. This tangential horizontal compression is thought, by its advocates, to be responsible for the upwards vertical movements of horsts, plateaus and geanticlines, and for the downwards movements of grabens, basins and geosynclines; as well as for folding, overthrusting and other deformations on a smaller scale.

Gravitational tectogenesis is not necessarily restricted to the epidermis of young, plastic sediments, which have been raised by orogenic forces; even though, of course, this situation is very favourable for the occurrence of gliding. The crystalline basement (the derm), can also collapse and glide under the influence of gravity.

Under the influence of internal volcanism, the deeper parts of the earth's crust or bathyderm, may also temporarily possess a smaller resistance against deformation. The material may therefore suffer more or less plastic deformations under the influence of stresses set up by the relief. Finally, it is also possible that mass displacements exclusively occur in the magma below the crystalline crust.

The types of gravitational tectogenesis which can be distinguished are shown in Fig. 21.

Before we discuss a few Indonesian examples of these types of gravitational tectogenesis, the question whether or not such gravitational reactions are possible, will be shortly discussed. Rocks

95

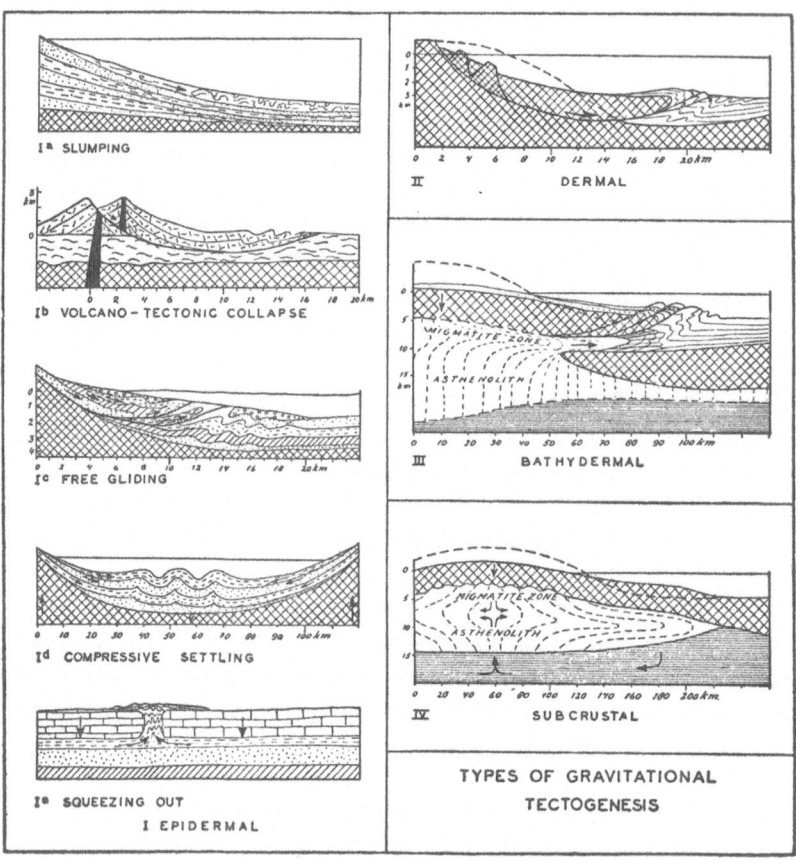

Fig. 21

Types of gravitational tectogenesis.

and mountains are, for man, symbols of strength and eternity. Also it is true that the stresses, set up by the differences in altitude, are comparatively small, especially when the high and low areas are rather remote and the average slope therefore small.

We must, however, realize, that these deformations take place on a time-scale which is entirely different from the scale we use in our direct surroundings and for our technical experiments. We cannot see molecules, nor can we see the thermal vibrations; yet there are numerous physical experiments, which give us a picture of the microcosm, entirely different from that based upon our direct sensory observation.

96

On the other side, in the macrocosm, we are not able to observe the movements of the earth's crust, as they occur on a time-scale, compared with which a lifetime is quite insignificant. If we could observe the changes, which take place in the course of tens of millions of years, on a human time-scale of minutes, we should be able to see the wave-like movements and the more or less plastic deformations of the apparently unchanging crust. Scientists who have made hydrodynamical experiments on a reduced scale, have thoroughly investigated the effect of this reduction. It follows from their studies, that one cannot apply laboratory values for the strength of various rocks to the geological conditions of growing mountains and subsiding basins.

If one imitates the deformation of a mountain chain, which in nature has taken one million years, by an experiment lasting one year, the time-scale has been reduced by a factor of 10^6. Calculation shows that the experiment should be done on a material of which the viscosity in 10^{18} times smaller than that of the rocks of which the mountain chain is made. This means, that only liquids of low viscosity can be used, such as solutions of gelatin (or diluted toothpaste)[1]. This material begins to flow on exceedingly small slopes.

An indirect result of this conception is the following:

When we draw a cross-section of a comparatively small object, e.g. a volcanic cone, it is possible to distinguish the various strata and structures even if we use equal vertical and horizontal scales. If such a volcano has been subjected to gravitational tectogenesis, the section still gives a satisfactory picture of the deformations which have taken place. If, however, we draw sections of areas with diameters of hundreds of kilometres, the topography and the thickness of the sedimentary strata have become so small, relative to the horizontal distances, that they can no longer be drawn on the same scale. When a cros-section of 200 kilometres has been reduced to 20 centimetres (1 : 1,000,000), differences of height of 1 km only amount to 1 mm. The vertical scale has to be exaggerated five times or more, in order to represent sedimentary series and structures. The result is, that the slopes in the drawing are too steep. The objection has therefore been made, that these drawings are too suggestive of gravitational tectogenesis. This now is not so; if we are

---

[1] M. GIGNOUX: *Méditations sur la tectonique d'écoulement par gravité*. Travaux Lab. géol. de l'Université de Grenoble. XXVII 1948, p. 12.

going to judge tectonic structures using our human conceptions of the strength of materials, it is better to increase the slopes in order to compensate the effect of the reduction in scale.

The larger the horizontal extent of a structural element, the smaller the "tectonic slope" it can withstand. A mountain-side of solid rocks remains undeformed, even with a slope of several tens of degrees, but a mountain as a whole can only stand an average slope of a few degrees. And of course, this means only that the mountain is stable according to human standards of time.

The peaks of the Nassau Mountains of New Guinea (the Carstensz peaks, 5030 metres) are higher than the limit of perennial snow in the tropics. This height, however, is geologically speaking only of a temporary nature. The present situation is only one still from a moving picture. The Snow Mountains are at present being pushed up by forces inside the earth; they are still growing, whereas the adjacent country, the Digul depression, is a marshy subsiding area. The southern flank of the Nassau or Snow Mountains is locally very steep. The rocks tend to slip down towards the Digul depression, and the entire mountain range is subjected to shear stresses, the trajectories of which are directed towards this depression.

Erosion is energetically attacking this growing range, and has already dissected it by steep valleys with furious mountain streams, which transport the debris to the foot of the mountain. It must be expected, however, that this destruction by erosion, as far as the tonnage of the material is concerned, will be finally exceeded by the removal of material by one or more of the types of gravitational tectogenesis.

Gravitational tectogenesis is caused by forces inside the earth; it is a reaction to the topographic relief created by the primary tectogenesis. Erosion and creep also transport material by using gravity, but as these processes are strongly dependent on external forces (climate) they are not considered as part of the gravitational tectogenesis. Landslides and similar phenomena constitute a transition from erosional to tectonic denudation. The velocity of landslides may amount to several metres per hour (or more) but examples are also known where the movements are hardly perceptible to human observation. The collapse of volcanoes can take place with velocities, which vary from a few metres per day to several tens of centimetres per year. The other types of epidermal gravitational tectogenesis can develop velocities which vary from metres to millimetres per year.

Another difference between tectonic denudation and erosional denudation, is that under tectonic denudation the relative position of the rocks is more or less preserved, as in glaciers, whereas erosion entirely destroys the original stratigraphic relations. By studying the tectonic structures the geologist is therefore able to reconstruct the original situation.

I. *Epidermal gravitational tectogenesis in Indonesia* (cf. Figure 21).

I*a*. Landslides are frequent in Indonesia, as it is a region with young tectonic and volcanic relief. The mountain sides are often covered in scars of landslides. A good example of the transition from erosional to tectonic denudation, is the great landslide of the Raung volcano in eastern Java. A section of the western slope of this volcano slipped down, and the debris now extend over a distance of 60 kilometres around the base, forming a swarm of hundreds of hillocks composed of lava blocks which rise some tens of metres above the surrounding rice fields. (See Fig. 22).

I*b*. Volcano-tectonic collapse is also frequent, especially in Java, where the volcanoes are often situated on a soft foundation of unconsolidated Late Tertiary sediments. The weight of these volcanoes exceeds several thousands of millions of tons, and it is not difficult to see that such cones finally must collapse under their own weight. This causes tensional structures in the top area, such as grabens and normal faults; and compresses, folds and overthrusts the material near the foot.

The Tambakan ridge (more than 700 metres high), near Segalaherang, which forms the foothills of the Sunda volcanic complex north of Bandung, was pushed up by the collapse of this mountain group of more than 2,000 metres height. This presumably happened in old-historic times, approximately 2,000 years ago, when the Sundanese were already living in the country. The legend of the origin of the Bandung Lake originated with this great disaster. (See Fig. 18).

The difference in height between the Sunda volcanic complex and the Tambakan ridge, is at present 1,000–1,500 metres, and their distance amounts to approximately 15 kilometres. The strength of the rocks is apparently sufficiently large to withstand this residual difference in height and the stresses it causes. The effect of the col-

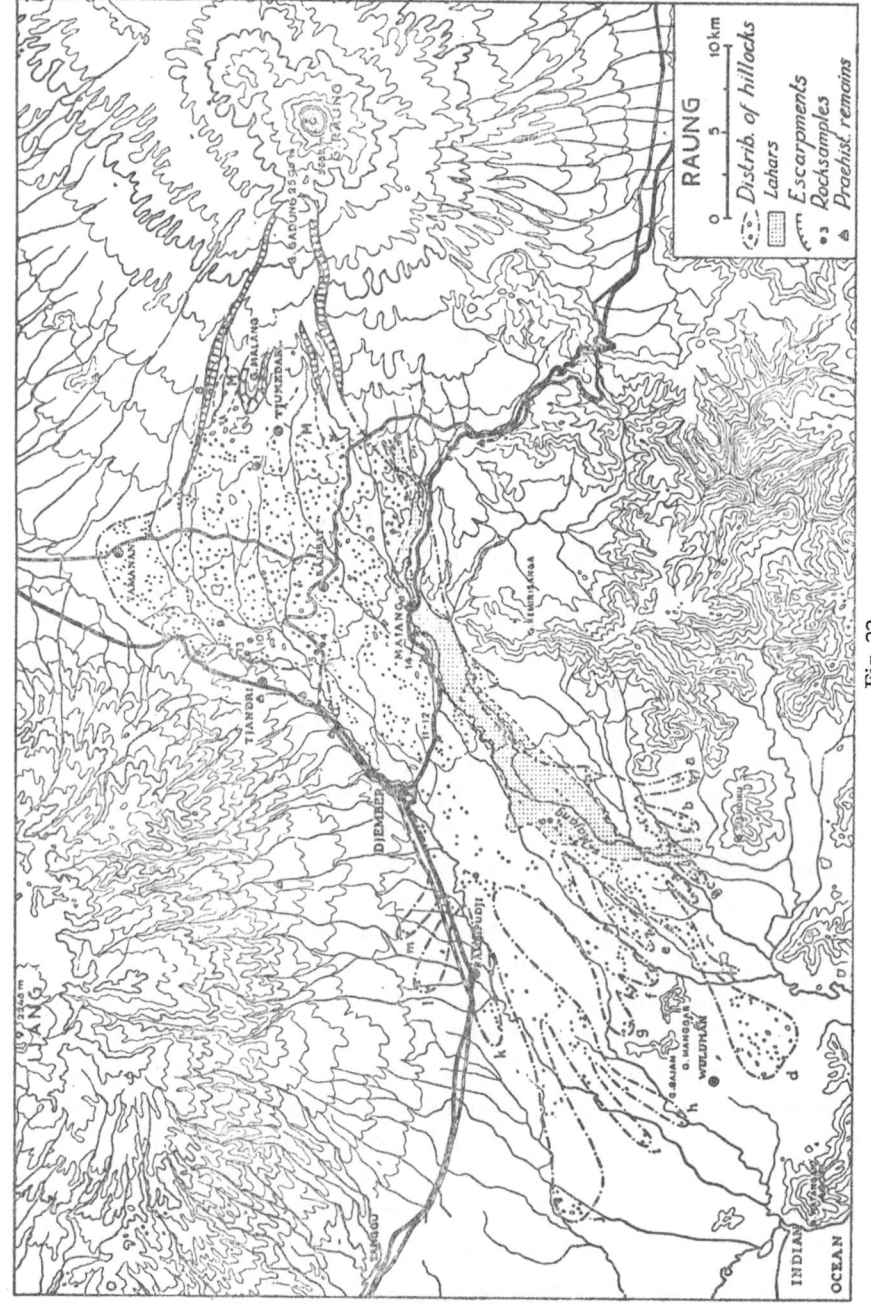

Fig. 22

A major slide of the western flank of Mt. Raung acc. to Neumann van Padang.

lapse was that the original height of this volcanic group decreased by a few hundreds of metres, and that from the top an area of approximately $7\frac{1}{2}$ km² was removed from the cross-section and added to that at the base.

Another good example of the collapse of a volcanic cone is the Merapi, in Middle Java. A part of this volcanic cone, which originally reached a height of approximately 3,300 metres, slipped in a westerly direction towards the Progo valley (Fig. 23). At a distance of 18–20 kilometres from the top, the basal strata of this volcano were pushed up against a shoulder of the Menoreh Mountains, and formed an anticlinorium, reaching a height of 452 metres. The collapse of this volcano was accompanied by a violent eruption, which must have taken place in the 10th century. This eruption put into disorder the Hindu-Javanese state in Central Java (which built the famous Borobudur and other temples) to such an extent, that more than five centuries elapsed before Central Java could again play a cultural rôle of any importance. The present-day steep and bare Merapi cone (altitude 2911 metres), is the result of nearly ten centuries of volcanic activity following this disaster. This new cone was built upon the ruins of the one dating from before the 10th century of the Christian era.

I*c*. Free gliding on mountain flanks can also be observed in Indonesia. Repeated triangulations have shown that two small volcanoes in middle Java, called Pawinihan and Telagalele, the latter of which still possesses a top crater, are drifting from their anchor, i.e. the eruption channel. They are now moving towards the Seraju valley near Bandjarnegara, with velocities of respectively 24 and 40 centimetres per year. The base of these volcanoes consists mainly of rather plastic Neogene marls and clays, which were pushed up into a mountain chain (the North Seraju Mountains) towards the end of Tertiary times. Magma broke through the culmination when the mountains were pushed up, and produced a number of volcanoes, which now form the Djembangan volcanic complex near Karangkobar. This complex of volcanoes has been, and is still gliding over the plastic base of clays and marls and moves partly to the north, partly to the south. (See Fig. 24 opp. to p. 49).

The underlying Neogene clays and marls also take part in this process of gliding. A tongue of this material has moved, in a glacier-like fashion, towards the Seraju valley along the southern flank of

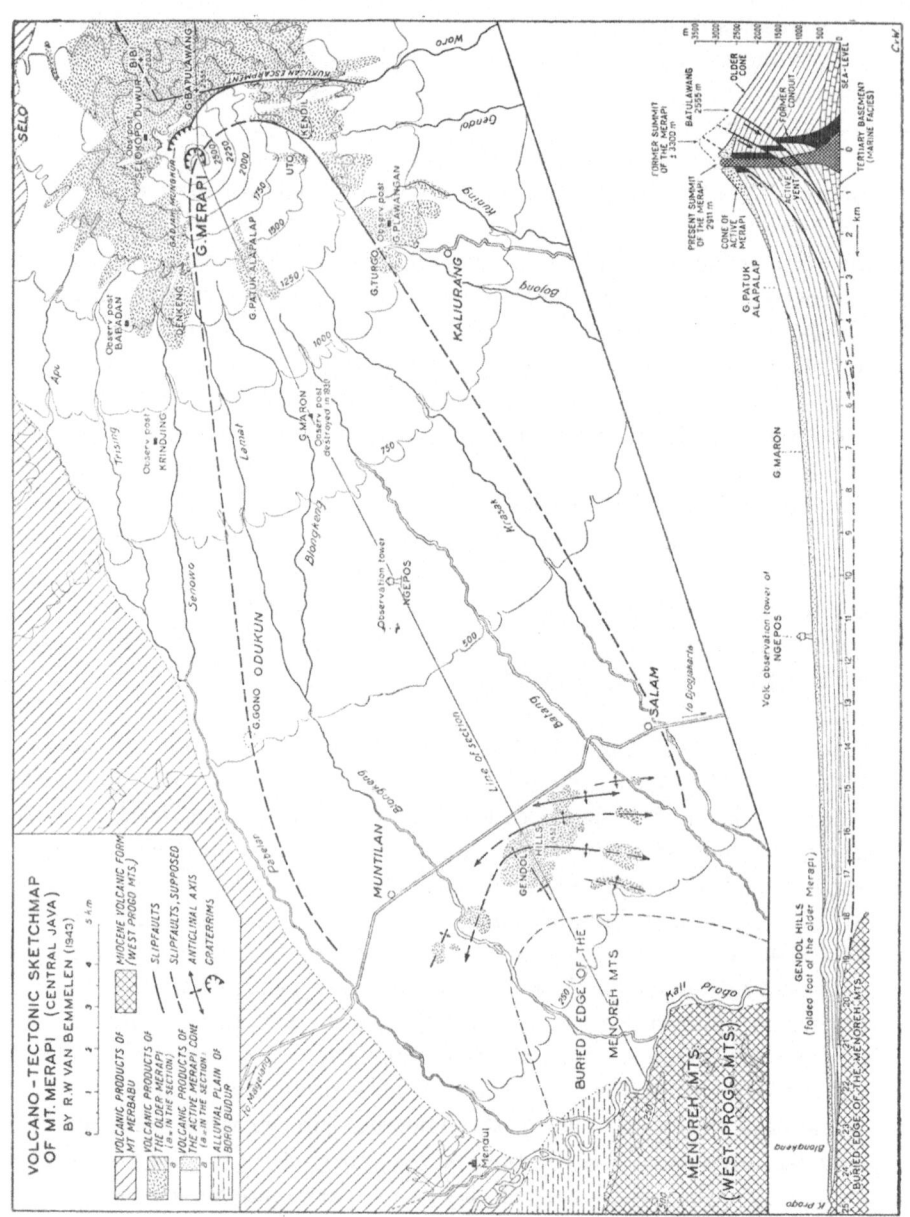

VOLCANO-TECTONIC SKETCHMAP
OF MT. MERAPI (CENTRAL JAVA)
BY R.W VAN BEMMELEN (1943)

0 1 2 3 4 5 km

VOLCANIC PRODUCTS OF
MT MERBABU

VOLCANIC PRODUCTS OF
THE OLDER MERAPI
(a. IN THE SECTION)

VOLCANIC PRODUCTS OF
THE ACTIVE MERAPI CONE
(a. IN THE SECTION)

ALLUVIAL PLAIN OF
BORO BUDUR

MIOCENE VOLCANIC FORM.
(WEST PROGO MTS)

SLIPFAULTS

SLIPFAULTS, SUPPOSED

ANTICLINAL AXIS

CRATERRIMS

102

younger volcanic deposits, and at present forms a "nappe" of Tertiary formations on younger, Pliocene and Quaternary volcanic agglomerates. The overlap amounts to approximately 10 kilometres.

Other nappes in Indonesia, such as those of Ceram, Timor and Djambi, with tectonic overlaps up to 100 kilometres or more, are presumably also due to gravitational gliding. These nappes are too thin and weak to be able to transfer a horizontal force acting from the roots to the front of the nappe, at the same time overcoming the friction at the base. Nor are there higher structural units which could have dragged these nappes along near their base. The overthrusts can only be explained by a force which acts on each particle, and transports the particles of the nappe without changing their relative position (see Fig. 25).

The Djambi nappes in southern Sumatra constitute an exceedingly difficult mechanical puzzle, as in this case the area of provenance of the nappes is situated at approximately 350 kilometres from their present position. In Chapter VI we shall endeavour to give a reconstruction of their journey.

I*d*. Compressive settling in subsiding basins can be studied in the idio-geosynclinal sedimentary basins, such as the Late Tertiary oil basins surrounding the Sunda Land. When sediments are deposited in a subsiding trough and the edges of the basin are formed by rising mountains (which supply the sediments) it may happen at a certain time that the process of erosion on one hand and of sedimentation on the other, does not wear down sufficiently quickly the tectonic relief which is being formed. Stresses are set up, the trajectories of which run from the higher to the lower areas, and also traverse the freshly deposited, still unconsolidated, sedimentary strata. The latter finally yield, and slump towards the deep part of the subsiding basin, where they are compressed into anticlinoria.

Sometimes these folds are concentrated in the central and originally deepest part of the basin as for example the Kendeng anticlinorium in eastern Java (Fig. 26); or they are concentrated along the edges of the rising bordering mountains, as for example along the eastern base of the mountains in Atjeh; or the axes of the folds are more or less regularly distributed over the entire width of the basin, as in Palembang and Djambi.

The folding movements are not necessarily restricted to the sedimentary contents of the basin; in many cases it appears that the

103

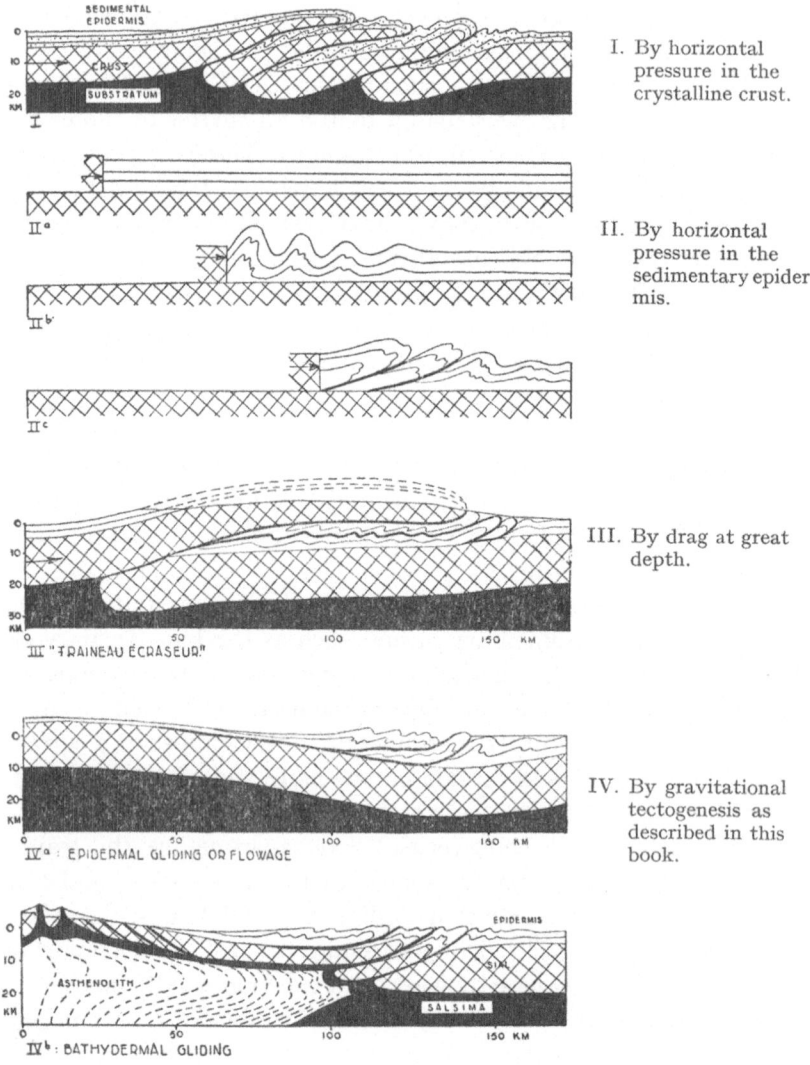

I. By horizontal pressure in the crystalline crust.

II. By horizontal pressure in the sedimentary epidermis.

III. By drag at great depth.

IV. By gravitational tectogenesis as described in this book.

Fig. 25

Theoretical possibilities for the origin of overthrusts.

104

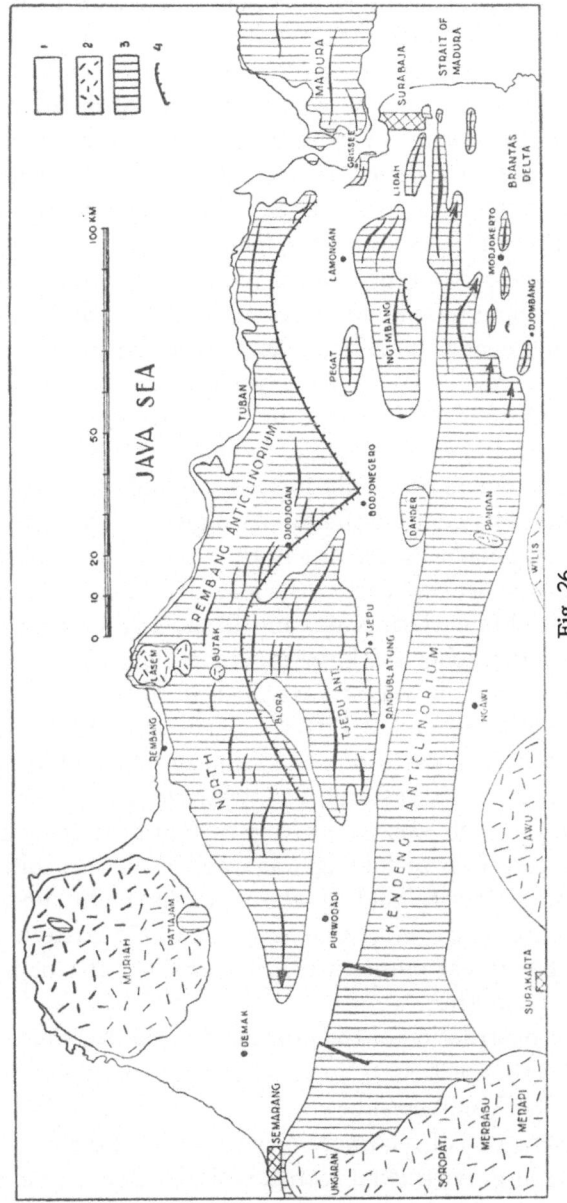

Fig. 26

Hypothetical faults and flexures in the basement complex of NE Java.
1. Alluvium. 2. Quaternary volcanoes. 3. Neogene . 4. Faults and flexures, partly buried by neogene
and quaternary deposits.

underlying basement has also taken part in it. These movements are therefore not always exclusively epidermal.

The shape of the folds depends on the local stresses which were in existence when the folds originated, on the slope of the basement on which the sediments were deposited and further on the thickness and composition of the sedimentary column and the mechanical properties of the stratigraphical units. This accounts for local changes in direction and intensity of the folding.

After compressive settling had taken place, orogenesis occurred in many cases in these mobile, unstable areas of the earth's crust. Primary tectogenetic movements took place, which entirely changed the original tectonic relief responsible for the compressive settling.

Anticlinoria, which were formed in a subsiding basin, may later be lifted up by orogenic forces and form a mountain ridge, and the adjacent areas, which originally formed the highest areas, may sink under sea level to a depth of thousands of metres. This inversion of tectonic relief is shown by eastern Borneo and Strait Macassar. In Tertiary times the deepest part of the idio-geosyncline of eastern Borneo was situated in eastern Kutai and Pasir. The Tertiary sedimentary column here reached thicknesses of more than 10 kilometres. This sedimentary column was compressed into the Samarinda anticlinorium towards the end of Tertiary times. This anticlinorium originated in the deeper parts of the basin, and the intensity of folding decreases in both easterly and westerly directions. The Macassar Sea, to the east, was a relatively high area at the time of folding, and the direction of folding was away from this zone and towards the west. After the folding, the Samarinda anticlinorium was pushed up into a low mountain range, but the eastern high area subsided and forms at present the Macassar deep, the depth of which exceeds 2,000 metres.

An accurate reconstruction of the orogenic relief at the time of the movements is therefore necessary for a proper understanding of the secondary tectogenetic movements.

I*e*. The squeezing out of relatively plastic strata also constitutes a type of epidermal gravitational tectogenesis. Plastic layers, such as gypsum, salt, clays and soft marls, may be squeezed out from between massive and more competent layers, such as limestone or sandstone beds and volcanic formations, and may be elsewhere injected into other strata, rise and finally extrude at the surface. In that case they behave like intrusive and extrusive magmas, of which the movements are also controlled by hydraulic principles. This is illustrated by the plastic Eocene clays and marls of Nanggulan, on the flank of the West Progo Mountains, which are squeezed from under the overburden of Oligocene volcanic formations, as it were like toothpaste from a tube. (Fig. 27).

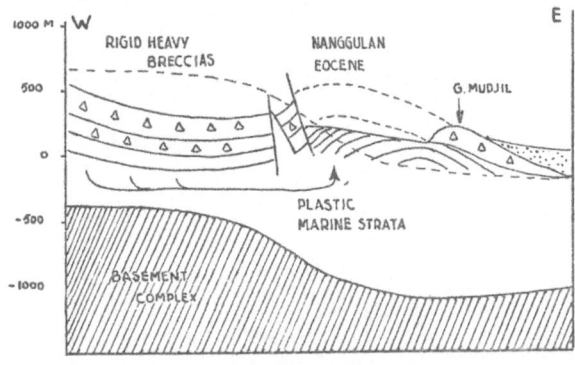

Fig. 27

Squeezing out of plastic marine strata on the eastern slope of the West Progo Mountains.

II. *Dermal gravitational tectogenesis*

The source of energy for, or the driving force of epidermal deformation, is provided by the potential energy acquired by the sediments during orogenic uplifts. This can often be made acceptable by geological and mechanical considerations. The principle of gravitational tectogenesis has therefore at present been fairly generally accepted for the relatively superficial movements of sedimentary series. The source of energy for the deformation of the crystalline basement is less obvious. For a long time endeavours have been made to explain the movements by hypothetical tangential compressive forces in the crust. This unicausal principle meets with many difficulties, now that our geological knowledge has so much increased, whereas the bicausal interpretation of earth movements, if applied in the right manner, can provide a harmonious picture of the structural evolution.

In the previous chapter we stated that the uplift of a geanticline is not necessarily the result of a lateral compression of the earth's crust, but that it may be due to magmatic forces pushing from below.

The increasing dimensions of a geanticline which is being

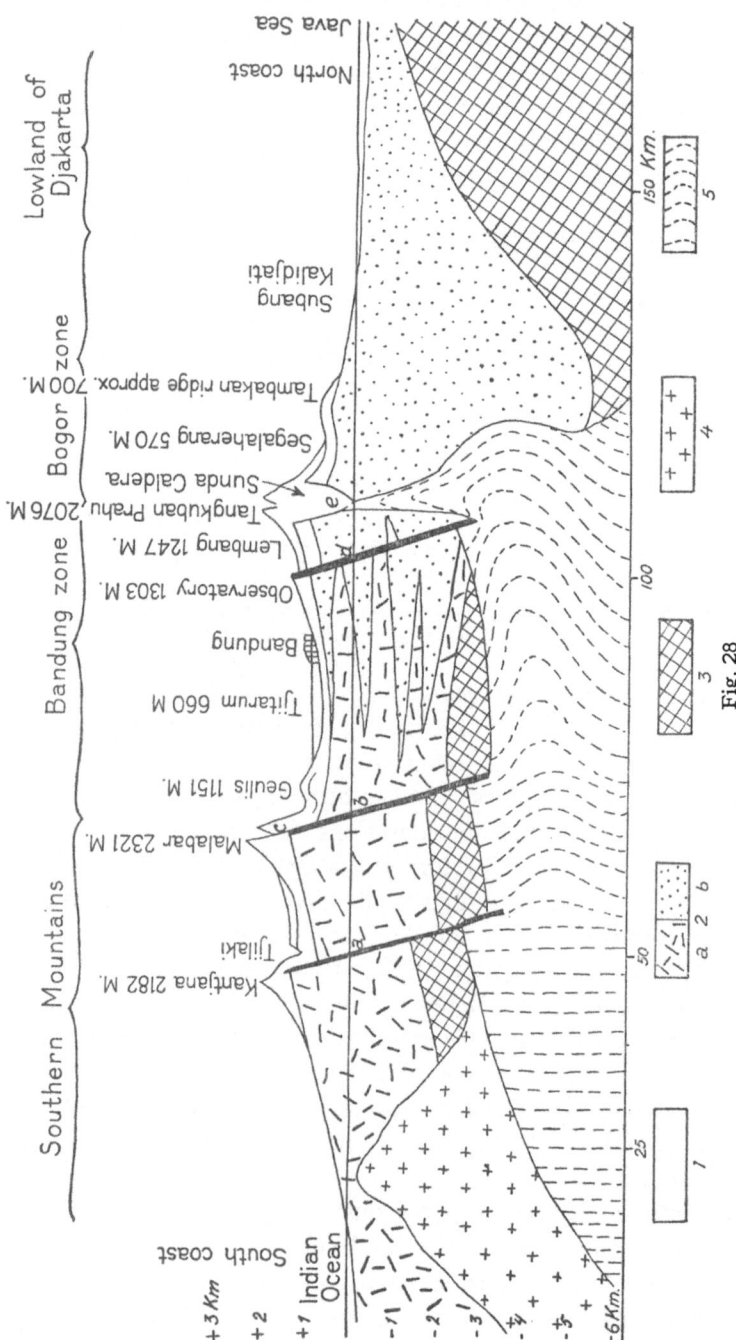

Fig. 28

Schematic section across West Java.

1. Quaternary terrestrial volcanic formation. 2. Tertiary volcanic and marine sediments. 3. Pre-tertiary basement complex. 4. Intra-miocene intrusion of granodiorite. 5. Plio-pleistocene level of migma and magma. The faults a–d originated respectively in Plio-Pleisto-cene, Old Pleistocene, Young Pleistocene and subrecent time. The fault e originated in historical time.

pushed up, gives rise to increasing internal stresses which in the long run can no longer be supported by the crustal rocks. This is illustrated by the geanticline of Sumatra, Java and the Lesser Sunda Islands. The geanticlinal structure collapsed and broke up into blocks which moved towards the adjacent subsiding strips.

In Flores a crustal block of 100 kilometres length and 30 kilometres width began to slide towards the deep Flores basin (-5,000 metres), and gave birth to the Maumere Bay. In Sumbawa a similar collapse of the geanticline gave birth to the Saleh Bay (see Fig. 20). Still farther west, along the entire length of Lombok, Bali and Java as far as the Wijnkoop Bay (a total distance of 1150 kilometres), the culmination of the geanticline became detached from the southern flank and slid away in a northerly direction (Fig. 18).

Fig. 28 indicates diagrammatically how the strength of the crystalline crust in the geanticline was reduced by corrosion from below, this being the result of rising volcanic matter. At the surface, the crust is covered by a sedimentary epidermis of several kilometres thickness which is particularly thick in the idio-geosyclinal basin of northern Java, where it consists for the greater part of plastic clays and marls. The epidermis of the culmination and the southern flank consist of more solid material of volcanic origin (intrusions and extrusions). Knowing this situation, it can be easily understood why this geanticline had to collaps, when the geanticline of Java was pushed up in Plio-Pleistocene times. Blocks of the top area (the Bandung zone amongst others) started sliding towards the geosyncline of northern Java and caused compression and folding of the Tertiary sediments (in the Bogor zone).

We are therefore dealing with tensional phenomena in an uplifted area, which are compensated by compression phenomena in the adjacent lower region; this is consequently a typical case of secondary tectogenesis.

Formerly it was thought, that tangential compression in the earth's crust was responsible for both the uplift of the geanticline and the folding in the sedimentary epidermis. Investigations in the Surakarta region in middle Java, where a detailed Quaternary stratigraphy, based on vertebrates, is available, have shown that the above-mentioned bicausal interpretation of earth movements is more likely. Many other instances from the Indonesian island arcs point to the same conclusion.

III. *Bathydermal gravitational tectogenesis*

The bathydermis is often brought into a mobile condition by volcanic processes. Magma and its emanations, coming from the depth of the earth, particularly during orogenesis, intrude the crust and transform part of the rocks into magmas of mixed origin, which are called migma. This is why the bathydermal zone is also called the migma- or migmatite zone. When the deeper parts of the crystalline crust, during orogenesis, have been mobilized into a plastic migma, and a geanticlinal ridge collapses, the stresses are transferred to this migma. It is then squeezed out towards the adjacent lower areas, and may there intrude the crust as palingenic (reborn) magma. These ideas on migmatization and the transformation of crustal rocks into palingenic magmas, have, especially in the last twenty years, come more to the foreground. We shall not discuss these ideas in any more detail at present, but only point out, that the movements of the material in the active migmatite zone, are in principle the same as those of the material squeezed out in the epidermal group. The scale of the movements in the bathydermal group is of course much larger.

During orogeny, the migmatite zone is situated at several kilometres depth in the crust, and it is only exposed in deeply eroded mountains. Migmatites have been described by ZEYLMANS VAN EMMICHOVEN from the Schwaner Mountains in central Borneo, but are, of course, best known from the older crystalline basements outside Indonesia.

When bathydermal gravitational tectogenesis takes place, the only thing we observe at the surface is that the uplifted area subsides again, and that the adjacent lower region rises. The lateral transport of material takes place at some depth, that is, either in the bathydermal migmatite zone, or in the partly still deeper magma zone.

The Sunda Straits area, between Java and Sumatra, provides an excellent illustration of such a bathydermal gravitational reaction to the orogenic relief. (fig. 29).

In Quaternary times the present position of the Sunda Straits was occupied by a crustal tumor, forming the most southern part of the Barisan geanticline. This tumor finally collapsed. The collapse was accompanied by paroxysmal eruptions of pumice, resembling granites and granodiorites in composition. The Krakatau eruption of 1883 is a last expression of this volcanism.

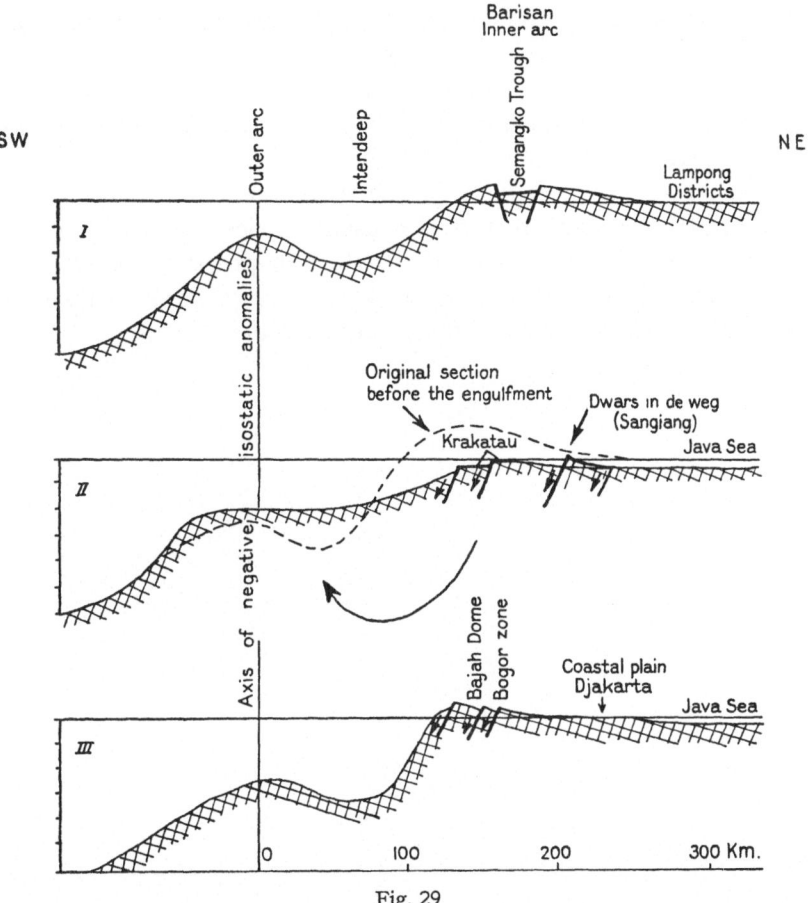

Fig. 29

Three sections across Strait Sunda and the neighbouring parts of Sumatra and Java.

During this collapse, migmatic and magmatic material was pushed away towards the interdeep. This interdeep was consequently pushed up from below and locally ceased to exist as such. This is the reason why it is not present in Profile II of figure 29. Instead of the interdeep, which is clearly visible in profiles I (southern Sumatra) and III (western Java), between the non-volcanic outer arc and the volcanic inner arc of the Sunda Mountain System, profile II shows an uninterrupted slope from the floor of the Sunda Straits to the foredeep. Along this slope the depth increases by more than 6,000 metres over a horizontal distance of 200 kilometres.

111

It appears from the evolutionary history of the Indonesian mountain ranges, that the orogenic phases, which create large differences in altitude between adjacent regions, are invariably followed by periods during which the relief decreases, as it were, by the flowage of the orogenic relief.

In chapter VI we shall sketch the evolution of the Sunda region, interpreted according to this bi-causality concept of tectogenesis. Before we do this, we shall first devote a chapter to the significance of geophysical observations for our knowledge concerning the mass distribution and structure of the deeper parts of the earth.

V. GEOPHYSICS

What does geophysics teach us?
We have interpreted volcanism and tectonics as the result of the transfer of internal energy to the surface. These processes leave traces in the form of chemical changes and mechanical deformations of the crustal rocks. These traces are studied by the geologist, who tries to reconstruct the evolution of a given area.

There is, however, another important source of information: geophysics. Geophysics comprises the study of gravity, of earthquakes, of terrestrial magnetism, etc.

There is a fundamental difference between geophysics and the specifically geological sources of information. The geophysicist studies the phenomena, which accompany volcanism and tectonics. He observes these phenomena as they are at the moment of observation. The element of time is therefore lacking in these observations. They do not teach us how gravity and seismic activity changed during the process of evolution. And that is something of great interest for geology, which is a typical historical science.

It is possible, of course, to frame hypotheses concerning the changes of the gravity field in time, or the changes in seismicity, etc., by applying and extrapolating the general laws of nature. This, however involves the risk to which all extrapolations are exposed; i.e. have we made sufficient allowance for all relevant factors?

When reconstructing the evolution of a given area, the geologist also has recourse to extrapolations. But he is guided by the direct observation of the traces, which the evolution has left in the earth's crust. He is helped and directed by the information which historical geology can give.

The geophysical information is nevertheless of great importance for the geologist. They provide him with information on the present condition of the earth as a whole or of particular areas. Gravity anomalies, earthquakes, and changes of the magnetic field, are phenomena which accompany the process of evolution. They provide important indications concerning the state of development which we observe at present. Applying LYELL's principle of uniformitarianism, which states that in the past the same laws and conditions prevailed as we observe at present, it is possible to obtain an idea of the geophysical phenomena which accompanied the evolution in the past.

What does gravimetry teach us?
If all shells of the earth had uniform density, and a shape perfectly adapted to the rotation of the earth, the value of gravity at any place on the surface could be derived from a general formula. Direct measurements of gravity have shown, that such a general formula in fact exists, but that nevertheless important deviations are found, which are called gravity anomalies.

We would expect anomalies to be caused by the topography. Continents which lie above sea level or mountains and oceanic basins and troughs, represent deviations from the ideal mass distribution. We would therefore expect deviations from the normal value of gravity according to the standard formula. It is remarkable that in general this is not the case. Gravity is approximately normal everywhere. This can only be explained by the hypothesis, that the surplus of mass represented by a continent or mountain, is compensated in the depths by the presence of material of a relatively small density. This has led to the hypothesis of isostasy or floating equilibrium, — in much the same way as an iceberg floats in sea water.

There is an infinite number of mathematical solutions for the distribution of mass, which causes a given gravity field at the surface. It can only be judged by geological, seismological and other considerations, which of these solutions best represents the reality. The question of how isostatic compensation comes about, is one of the most difficult problems in geophysics.

The study of earthquakes shows, that in the outer 100–200 kilometres of the silicate shell of the earth — i.e., the tectosphere — several layers occur. These layers are not everywhere of the same thickness nor are they everywhere present. The structure below the Pacific Ocean is probably different from that under a continent, and again different from that under a mountain system in the process of formation. The processes which occur in the tectosphere during the geological evolution, cause temporary and local changes in the thickness, physical state, chemical composition and temperature of the various layèrs. All this influences the gravity field at the surface.

The processes which occur in the depths tend in general towards equilibrium, that is to say not only towards isostatic or floating equilibrium, but towards perfect hydrostatic equilibrium, where the shells have a uniform density, are everywhere equally thick and adapted to the rotation of the earth.

This equilibrium has still not been reached, in spite off three thousand million years of geological evolution. This is so, because the earth possesses large stores of internal energy, which have not yet been exhausted. This internal energy can take various forms, such as heat, chemical energy and deviations from the gravitational equilibrium. In the course of the evolution this energy is gradually released and complicated chain reactions occur in the depth. These reactions make themselves felt at the surface as volcanic and tectonic processes, and are also accompanied by geophysical phenomena, such as changes in the gravity field, in the magnetic field, and as earthquakes.

Large gravity anomalies are found in Indonesia, as orogenesis is still in full swing at many places in the archipelago.

A great number of gravity measurements have been made in Indonesia; on land mainly by oil companies, and at sea by VENING MEINESZ on board submarines. Investigations have been made in order to see in how far these measurements are in accordance with the normal values and with the hypothesis of floating equilibrium of the crust. The differences between measured and calculated values are called isostatic anomalies. The lines of equal anomaly are called isogams. An isogam map for the entire archipelago, based on the hypothesis of regional isostatic compensation as developed by VENING MEINESZ, has been published by DE BRUYN in 1951[1]. On this map a zone of pronounced negative isostatic anomalies can be seen, which stretches from the Arakan Yoma chain in Burma, across the Andaman and Nicobar Islands and the islands west of Sumatra to the submarine ridge south of Java, Bali and Lombok. After a break near

---

[1] J. W. DE BRUYN. "Isogam Maps of Caribbean Sea and Surroundings and of South East Asia". Proc. IIIrd World Petroleum Congress, the Hague, 1951, pp. 598–612.

Sumba this zone continues across Sawu, Roti, Timor, and the Babar, Tanimbar and Kai Islands to Ceram. Near Ceram this zone splits up into two zones, one of which crosses the Ceram Sea in a north-westerly direction and continues across Obi-major towards the Moluccan Sea, where it joins the second zone. This second zone forms a great loop, which runs across Buru, the Tukang Besi Islands, Buton, Muna, and the south-eastern and eastern peninsulas of Celebes towards the Moluccan Sea. In the Moluccan Sea negative isostatic anomalies have been found which belong to the largest known (— 205 milligal). They disappear when the zone reaches the Philippine deep.

This zone of negative isostatic anomalies (total length approximately 8,000 kilometres), was discovered by VENING MEINESZ during his pioneer work on board submarines (1923, 1926, 1929–30) and is therefore often called the "Vening Meinesz zone".

It is a striking fact, that the Vening Meinesz zone for the greater part coincides with young, non-volcanic mountain chains, which for the first time in their history are being pushed up from the geosynclinal basins. These ridges are mountain chains in the process of formation, and are at present still slowly rising. Over large distances they only form submarine ridges which have been pushed up from the deeper part of an oceanic basin, as for example south of Java. Elsewhere only a few islands appear above sea level, as for example west of Sumatra. Raised coral reefs and other phenomena prove their recent uplift. These young mountain chains form the non-volcanic outer arcs of mountain systems. The Sunda Mountain System is along its entire length accompanied by a non-volcanic outer arc, characterized by negative isostatic anomalies. The non-volcanic outer arc of the Celebes Mountain System also coincides with the Vening Meinesz zone. In the Moluccan Sea the non-volcanic outer arcs of two mountain systems, which face each other, coincide. The volcanic inner arcs of these two systems form the row of volcanoes of the Minahassa and Sangihe Islands to the west, and the row of volcanoes of Halmaheira and Ternate east of the Moluccan Sea. The complicated, predominantly submarine system of ridges in the Moluccan Sea, on which the islands of Maju and Tifore are situated, forms the common outer arc of these two mountain systems. It is remarkable that here the negative isostatic anomalies are nearly twice as large as in other sections of the Vening Meinesz zone.

The branch of the Vening Meinesz zone which runs from Ceram to the Moluccan Sea, does not coincide with a geanticlinal structure. This branch is as it were a short cut across the large loop across Celebes.

Next to the Vening Meinesz zone strips and larger areas of positive isostatic anomalies occur. It is remarkable, that the largest positive and negative isostatic anomalies are situated in areas where at present orogenesis is most active, as for example in the outer arc of the Sunda Mountain System and in the Moluccas. In the central Sunda Land, which is more stable, the anomalies are much smaller. It is further remarkable, that the negative anomalies coincide with areas which tend to rise; whereas the strongly positive fields occur in young subsiding basins. The latter relation may be observed in the Banda, Celebes and the Sulu basins, the Macassar deep and the trough between the southern and south-eastern peninsula of Celebes.

In addition to the Vening Meinesz zone, many other strips of negative anomalies can be distinguished, though they are of smaller size and intensity. The most pronounced one is the zone of negative anomalies, which coincides with the Kendeng ridge and the Madura Straits near eastern Java. Just as the outer arc, the Kendeng zone is a geanticline in the process of formation. Here too a relation seems to exist between the presence of negative anomalies and tectonic uplift, in the same way as a relation seems to exist between positive anomalies and subsidence.

The most obvious interpretation of this relation is that the rock column which underlies the zone of negative anomalies is too light relative to its surroundings and is therefore pushed up; whereas the rock column underneath the subsiding basins is too heavy and therefore tends to sink.

The oil companies have executed gravimetrical investigations in restricted areas in order to determine the mass distribution near the surface. This is an important aid in finding structures favourable for the accumulation of oil.

Geophysicists are more interested in the regional gravity field, both on land and at sea. This provides information on the mass distribution at greater depth in and below the earth's crust. Both the measurements on land and at sea are used for regional studies.

The best founded interpretation of the regional gravity field is VENING MEINESZ's hypothesis that the zone of negative anomalies is

117

caused by a root of granitic material at the bottom of the crust. This root is pushed up by the surrounding material of greater density and causes a rising ridge at the surfaces which for example forms the islands of the outer arc of the Sunda Mountain System.

The origin of such roots of mountains can be explained in two fundamentally different ways. VENING MEINESZ has suggested that the earth's crust is buckled and its granitic material pushed into the substratum. The present author assumes the existence of an accumulation of granitic magma at the bottom of the crust, caused by chemical processes. This magmatic root is called an "asthenolith"

What does seismology teach us?
The forces inside the earth cause tectonic deformations and volcanic transformations of the earth's crust. The accompanying mass displacement may take place so slowly and gradually as to be hardly perceptible at the surface. But it may also happen that stresses in the crust cause elastic deformations, which at a certain moment may result in the formation of a fault (rupture). This is observed at the surface as an earthquake. Earthquakes therefore occur when locally the internal stresses accumulate so quickly, that they can no longer be neutralized by more or less plastic deformations of the crustal rocks.

Both crystalline and glassy rocks in the outer part of the earth can break and cause earthquakes. It depends on the viscosity of the rocks and the velocity with which the stresses accumulate whether they will break or suffer plastic deformation; this can also be demonstrated with materials like pitch or sealing wax.

High temperatures and high hydrostatic pressure encourage plastic deformation. This is why the normal tectonic earthquakes tend to occur in the outer shells of the earth (maximum depth approximately 50 kilometres). Since 1928, however, earthquake foci have been determined, which are situated at a depth of 100 to 250 kilometres; and quakes have even been observed which are due to rupture at depths of 300 to 700 kilometres. These intermediate and deep earthquakes can only originate when the stresses accumulate very quickly. It is the factor of time and not of depth, nor that of the glassy or crystalline nature of the rock concerned, which determines whether plastic deformation or rupture will occur.

In addition to normal, intermediate and deep tectonic earthquakes,

118

volcanic earthquakes are distinguished as a separate class. The latter are restricted to external volcanism, their focus is always situated at a shallow depth and the mass which takes part in the movement, is mostly much less than in tectonic earthquakes. They are therefore nearly always comparatively weak and seldom perceptible for human beings, except in the direct neighbourhood of the volcano. The earthquakes which are generally felt are in nearly all cases of tectonic origin and have nothing to do with external volcanic eruptions. External volcanism and earthquakes nevertheless occur both in regions where earth movements are particularly powerful. In Indonesia both are found in areas where active orogenesis takes place, such as the Sunda Mountain System, the Moluccas and New Guinea, but not in the more stable regions such as the Sunda Land and the Sahul shelf.

The frequency of earthquakes is very high in areas of active orogenesis. The number of earthquakes in 1936 for various regions is listed below.

Sumatra	149
Java	128
Lesser Sunda Islands	27
Celebes	79
Moluccas	81
New Guinea	26
Total	490

Heavy earthquakes, causing large-scale destruction, are fortunately rather rare. Once in every four years in the entire Archipelago, which spans $\frac{1}{8}$ of the earth's circumference, a destructive earthquake occurs, affecting an area of more or less restricted size. The geologist is able to indicate areas where destructive earthquakes may occur. These are areas where active earth movements take place. Considerable protection against the destructive effects of earthquakes is provided by specially designed houses, for example wood-frame houses or houses of reinforced concrete.

During the large earthquake which occurred in Java on 23rd July 1943 and caused the death of more than 4,000 people and the destruction of more than 12,000 houses, the greatest losses were caused by the collapse of the badly built stone houses of the middle classes. A properly built house goes through an earthquake like a good ship weathers the storm. It is therefore necessary to build solid

119

houses in areas where earthquakes are frequent. This will reduce the loss of life to a minimum.

VISSER observed that for Java and Sumatra the greater part of the epicentres was situated under the sea along the edge of the Asiatic continent. For Sumatra the following figures were found for the period between 1909 and 1932.

Number of epicentres on land 204
Number of submarine epicentres . . . 1105
Total number of earthquakes 1309

For Java the following figures, referring to the period between 1909 and 1921, are available.

Number of epicentres on land 13
Number of submarine epicentres . . . 846
Total number of earthquakes 859

Certain zones are also found on land, where earthquakes are frequent. One example is the Semangko zone in Sumatra. It is a system of faults and grabens, which cut through the culmination of the Barisan geanticline along its entire length. In this Semangko zone (fig. 30), repeated tensional movements and relative displacements of fault blocks occur, accompanied by destructive earthquakes, such as those of Tapanuli (1892), Kerintji (1909), the Padang uplands (1926) and Liwa (1933).

After the Tapanuli earthquake in 1892, it was found that not only vertical, but also sudden horizontal displacements had occurred. The trigonometrical points on the volcanoes Malintang (1983 metres) and Sorik Marapi (2145 metres) and on the non-volcanic Manondong mountain (1244 metres), all situated on a ridge of 50 km length west of the Semangko fault, suffered a displacement

Fig. 30

The seismic belt in the Semangko zone of Sumatra.

of 1.2 to 1.3 metres in a northerly direction; whereas the Tor si Hite (1407 metres), situated 12 km east of this fault in the Semangko graben, moved 0.6 metres in a south-easterly direction.

These movements are the movements of smaller blocks, which take up new positions relative to each other during the shock. The general character of the movements along the Semangko zone is due to tension at right angles to the axis of the Barisan; this tension causes longitudinal faults, grabens, volcanic fissure eruptions and even open fissures, as for example during the Liwa earthquake in 1933.

Earthquakes remind us of the fact that even at present earth movements are still taking place. Mountains are being created or worn away, but we are not able to notice any changes during a lifetime. This is, however, not the most important aspect for the geologist. He is often able to prove the occurrence of earth movements by other means, such as repeated triangulations or by morphological studies (terraces, rejuvenated erosion, drowned coasts, etc.).

After the earthquake waves have passed through the earth, they are registered by very sensitive instruments, called seismographs. The records of these seismographs, called seismograms, enable one to determine very accurately the time of arrival of the various sorts of earthquake waves. The two most important ones are the condensation (or longitudinal, or primary) waves and the shear (or transverse, or secondary) waves, which have different velocities. The velocity depends on the elastic properties and the densities of the rocks.

The earthquake waves tell us something about the depth of the discontinuities inside the earth, as well as about the elastic properties of the material. In 1950, GUTENBERG[1] published figures for the various velocities of longitudinal waves (V_l) in the tectosphere. The present author proposes the following geological interpretation of these figures (fig. 31).

In the crystalline granitic crust, the velocity of longitudinal waves increases from 6 km/sec to $6\frac{3}{4}$–7 km/sec at a depth of approximately 10 km. At a depth of approx. 15 km, the velocity decreases again to a minimum of $5\frac{1}{2}$ km/sec at a depth of 20–25 km. This so-called seismic sofar channel coincides with the migmatite zone, where the rocks have been melted and mobilized and are less elastic than in the upper crystalline part of the crust. Deeper down the velocity

---

[1] Science, Vol. 111, 13 Jan. 1950, No. 2872, pp. 29–30.

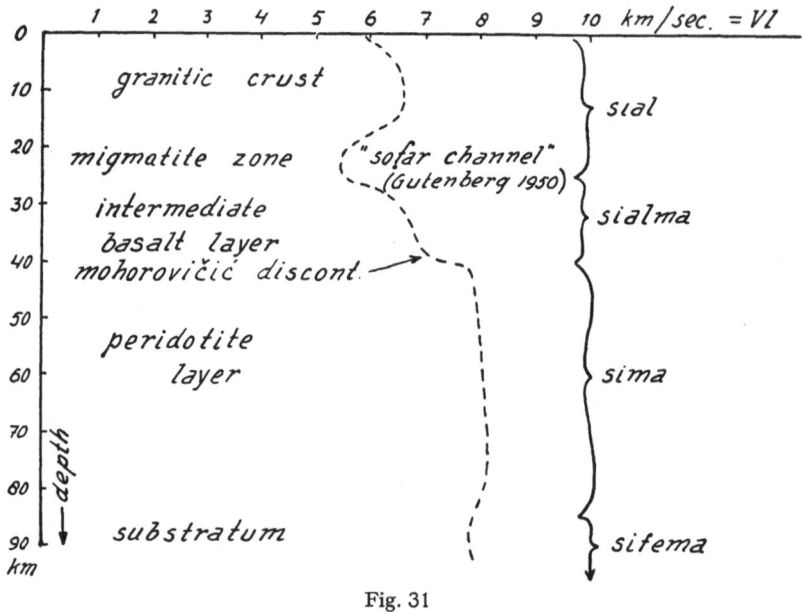

Fig. 31

Possible structure of the tectosphere on the strength of the velocity of condensation (longitudinal) waves.

of the longitudinal waves increases again, perhaps with a jump, to values of 6½–7 km/sec in the intermediate basaltic layer. At a depth of approximately 40 km (it varies between 30 and 60 km), another jump occurs and the velocity increases to 8.1–8.2 km/sec. This is the important Mohorovičić discontinuity, which separates the glassy intermediate basaltic layer from the predominantly crystalline shell of ultrabasic rocks. Some melting may also occur at the bottom of this crystalline layer, causing a decrease in elasticity and consequently in the velocity of the seismic waves. This may account for the decrease to 7.9 km/sec. at a depth of 80 to 150–180 km. Still deeper down, the velocity increases continuously to a depth of at least 900 km. This latter part of the earth is the actual substratum, in which presumably the iron and magnesium content increases and the silica content decreases with depth.

The earthquake waves therefore teach us that the tectosphere consists of several shells. Three principal shells at least can be distinguished, namely a crystalline crust of granitic composition, a presumably glassy intermediate layer of basaltic composition and a predominantly crystalline layer of ultrabasic composition. The two crystalline shells, the acid and the ultrabasic, are presumably sometimes corroded (melting, chemical mobilization) by the underlying vitreous shells. This decreases the elasticity and consequently the velocity of the seismic waves.

We mentioned at the beginning of this chapter, that the earthquake foci tend to be concentrated at three different depths. The normal tectonic earthquakes originate in the layers above the Mohorovičić discontinuity, the intermediate quakes seem to originate near the boundary between the crystalline ultrabasic zone and the vitreous substratum, and the deep quakes seem to occur in this glassy substratum.

The distribution of the normal, intermediate and deep earthquake foci shows a remarkable regularity, which in Indonesia was first described by BERLAGE. The depth of the foci increases with the distance from the open ocean. The epicentres of quakes at approx. 100 km depth roughly coincide with the oceanic coasts of Java and Sumatra, whereas the epicentres of quakes at 600–700 km depth below the Java and the Flores Sea are situated several hundreds of kilometres inland.

Figure 32 shows a cross section of the Sunda Land and the adjacent part of the Indian Ocean. It shows the layers of the tectosphere and the two arcs of the Sunda Mountain System, as well as the roots which support these arcs. A number of important normal, intermediate and deep earthquake foci have been plotted in their position relative to the axis of the inner arc. There are three more or less clearly defined clusters, of which the upper one is situated between the inner and the outer arc and above the Mohorovičić discontinuity, the intermediate one for the greater part under the volcanic inner arc and the lower one occurs below the inner side of the volcanic inner arc and farther away from the ocean. The foci are therefore arranged in a zone which dips away from the Sunda Mountain System towards the Sunda Land at an angle of circa 55°.

The normal earthquakes almost certainly accompany the orogenesis which we observe at the surface. This relation is not so clear for the intermediate and deep foci, even though their distribution indicates that a relation of some sort must exist. The suggestion has been made that the intermediate quakes should accompany the volcanic activity of the inner arc, as they occur here, as well as else-

123

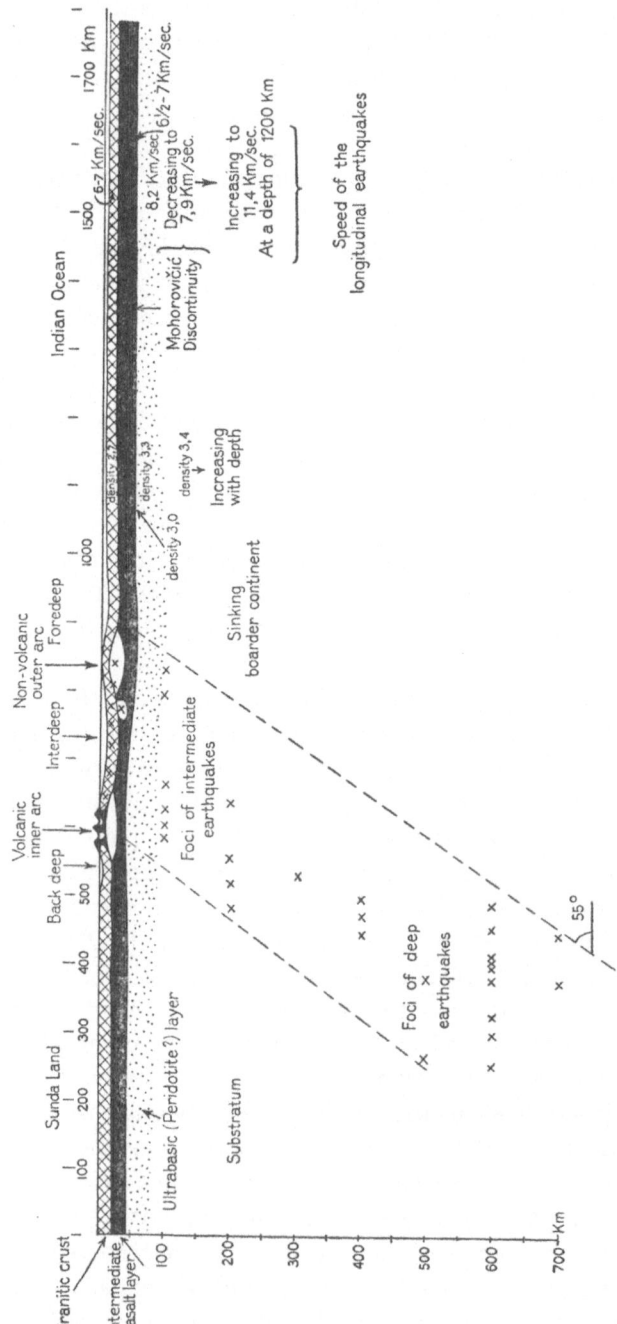

Fig. 32

Position of the normal, intermediate and deep seated earthquake foci with respect to the axis of the Sunda Mountain System.

124

where on earth, below it. Figure 32 however, shows that this relation can only be indirect and is the result of the landward shift of the intermediate foci, relative to the normal ones.

It seems probable that the intermediate and deep earhquakes are not directly connected with local orogenesis, as the associated mass displacements are not restricted to the tectosphere. A greater depth of the movements must cause tectonic phenomena of a greater regional extent. The intermediate and deep quakes are therefore thought to be associated with epeirogenic movements which extend over thousands of kilometres, rather than with orogenic movements which only cause deformations in zones 100–200 km wide.

It appears that such movements have taken place in and around the Archipelago in the youngest part of the geological history and presumably are still going on. The eastern part of the Archipelago has subsided relative to the Sunda Land and Australia. In former times the Archipelago was surrounded by borderlands, which have since then subsided to oceanic depths. Examples are Philippinia east of the Philippines, Northern Melanesia north of New Guinea, and Olim Gondwana in the Indian Ocean. It may be assumed that elsewhere on earth where intermediate and deep earthquakes occur, epeirogenic movements extending over large areas are taking place, and may even lead to the subsidence of continental regions below sea level.

KONING[1] and RITSEMA[2] have investigated the mechanism of the faulting in some deep earthquake foci. Their researches have established that deep earthquakes are caused by sudden normal faulting due to tensile stresses in the focus. These movements occurred along one of two mutually perpendicular planes, possessing a horizontal intersecting line. This line has the same bearing as the zone on which the normal, intermediate and deep earthquake foci are situated. One of the above mentioned planes is more or less parallel to this zone and dips away under the continent at an angle of 55°, the other plane is perpendicular to the first and dips away under the ocean at an angle of 35°.

The subsidence of borderlands in the Pacific and Indian Ocean and of the Moluccas region relative to Asia and New Guinea are

[1] L. P. G. KONING: Over het mechanisme in de haard van diepe aardbevingen. Dissertation Amsterdam 1941.

[2] A. R. RITSEMA: Over diepe aardbevingen in de Indische Archipel. Dissertation Utrecht 1952.

considered by the present author as due to physico-chemical processes which lead to changes in density. The voluminous basaltic intrusions and effusions, for example, after cooling must increase the average density of the crust. It becomes too heavy and begins to subside.

The physico-chemical processes which accompany mountain building must also lead to changes in density at greater depths below the surface. Parts of the silicate shell which have become too heavy in relation to their surroundings, tend to move downwards.

This downward movement of extensive silicate masses is in general brought about by plastic deformations, but apparently rupture occurs locally. It is not possible to decide from the observations whether this faulting takes place along the plane which dips away under the continent at 55° or along the plane dipping in the opposite direction at an angle of 35°. It is possible that both play a rôle, their relative importance depending on local conditions.

KONING [1], not only takes into account the position, but also the magnitudes of the deep earthquake foci. This may contribute towards a better understanding of this interesting phenomenon.

---

[1] L. P. G. KONING: "Earthquakes in relation to their geographical distribution, depth and magnitude".
Proc. Kon. Ned. Akad. v. Wetenschappen, Amsterdam, Vol. 55, Series B, nos. 1–3,1952.

VI. GEOLOGICAL EVOLUTION

General principles

In the chapter on stratigraphy we have discussed the chronology of the geological events and made mention of the methods which the stratigrapher applies.

In the chapters on volcanism and tectonics we have considered the chemical and mechanical aspects of the geological evolution.

In the chapter on geophysics we have discussed the phenomena which accompany this evolution at present, such as gravity anomalies and earthquakes, and we have investigated what these phenomena can teach us of the mass distribution and mass displacements at greater depths.

In this chapter we shall arrange the various expressions of the forces inside the earth in a scheme of space and time, in order to arrive at a synthesis of the evolution, which has created Indonesia.

When we discussed the distribution of volcanism in the course of the orogenic development, we demonstrated by examples that each structural zone passes through a number of phases, which are characterized by certain consanguineous igneous rocks. The process of orogeny begins with the subsidence of a narrow zone of the earth's crust. This zone forms the foredeep of another orogenic zone which is in a more advanced stage of evolution. This initial or geosynclinal phase of orogenesis is accompanied by ophiolitic volcanism.

After an incubation period of several tens of millions of years the first uplift of the sedimentary basin takes place, and the first proper phase of orogenic evolution is reached. This orogenic phase can be split up into a series of uplifts, separated by periods of decreasing relief and subsidence. Each of these intervals between the orogenic subphases lasts some several ten millions of years and the intensity of the uplifts quickly decreases. For most mountain chains in Indo-

127

nesia three periods of uplift can be distinguished; the first one is not volcanic, the second one is accompanied by external volcanism and the third one is no longer volcanic. In Sumatra, Java and the Philippines the third uplift is accompanied by a revival of the volcanic activity. It appears that the uplifts are always accompanied by the intrusion of granites into the core of the geanticline.

We have also seen, when we discussed palaeovolcanism, that each mountain system consists of a number of parallel, arcuate structural zones, of which the outer one is the youngest and the others represent successively older phases of evolution when moving towards the concave side of the system.

The result is that a mountain system migrates in lateral direction in the course of the orogenic development. It appears that foci or centres of diastrophism can be distinguished, from which orogenic cycles migrate in lateral direction as wavelike deformations of the earth's crust.

The Sunda region in western Indonesia is a good example of this orogenic development. Here, orogenesis began in Devonian times in a central zone which ran across the Anambas Islands and the Schwaner Mountains in Central Borneo towards the Telen region in eastern Borneo. From this focus of diastrophism an orogenic cycle migrated in northerly direction towards the Asiatic hinterland; another cycle of orogenesis migrated in southerly direction towards the Gondwana foreland (at present the Indian Ocean), as indicated in Fig. 33.

In the following paragraphs we shall discuss the evolution of the Sunda region in some detail.

The geological history of the Sunda region
In order to give a clear representation of the development of the Sunda region, one should have to show a moving picture, which reduces the time taken (250–300 millions of years) to a quarter of an hour. This comes to a reduction of approximately ten million million times. On this time scale we cannot observe waves or tides but instead we should see the earth's crust heaving like a sea. We do not possess sufficiently detailed informations for such a film, nor is it an easy task to draw the thousands of individual pictures. In the following pages we shall therefore exemplify the evolution by fourteen block diagrams, showing successive stages in the evolution.

128

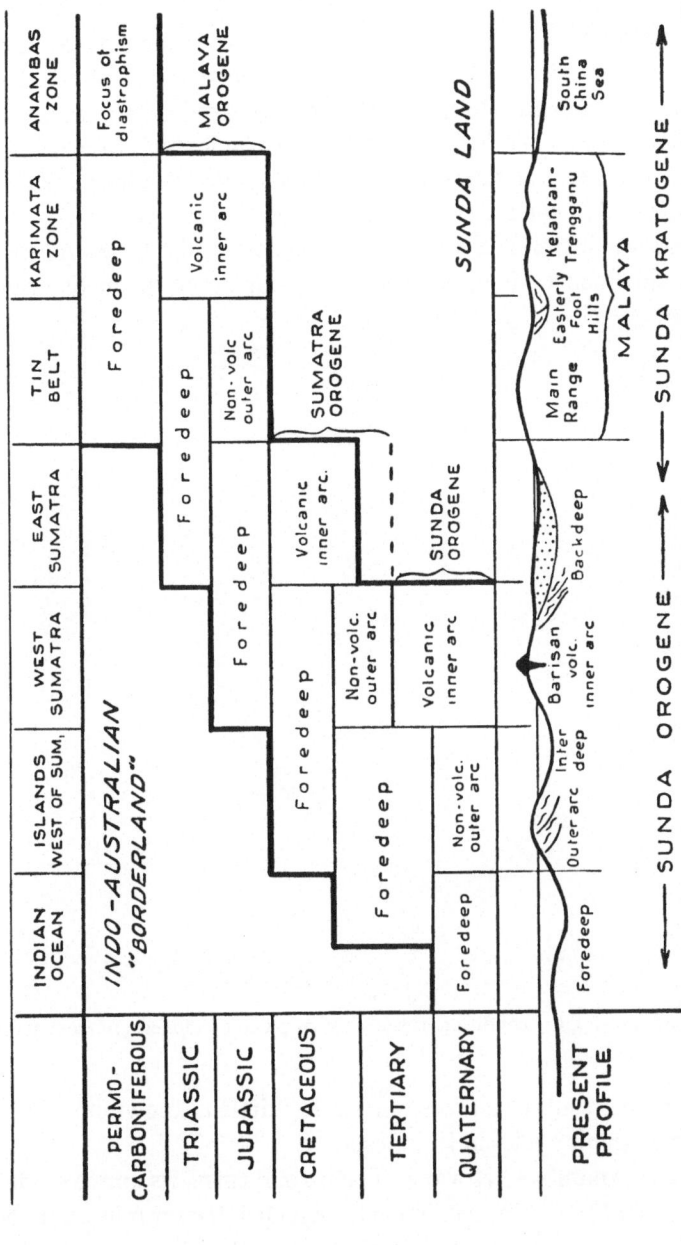

Fig. 33

The evolution of the mountain system in the Sunda region in time and space. In succession the Malaya orogene, the Sumatra orogene and the Sunda orogene originated, shifting each time more southwestward with respect to the starting point of the orogenesis.

Devonian

Fig. 34 represents the conditions in Devonian times, when the Indonesian Primeval Continent in the Anambas-Schwaner-Telen zone of the Sunda region broke up and disappeared under sea level. This subsidence was the beginning of an orogenic evolution, which has lasted hundreds of millions of years and the results of which we see at present in the Sunda region (western Indonesia).

Geological evolution is a long chain of causes and effects and this Devonian subsidence therefore did not occur spontaneously. This

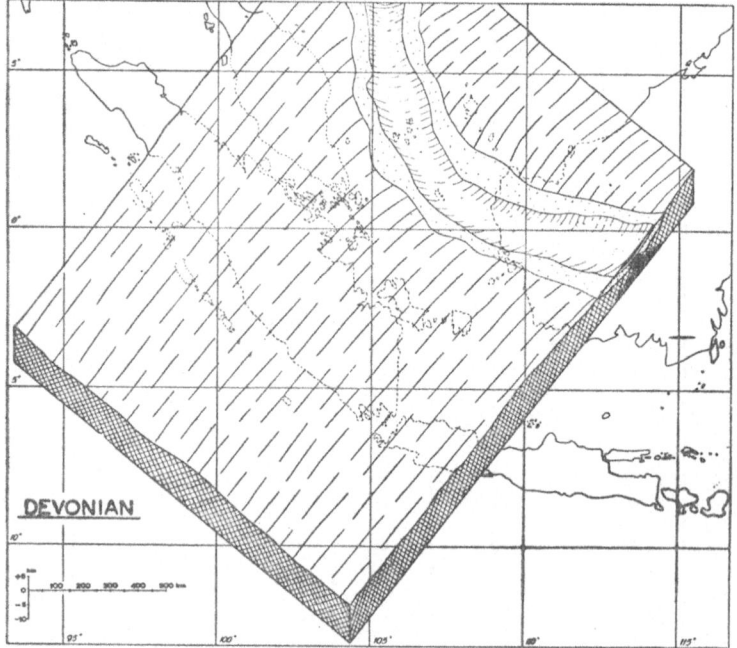

Fig. 34

Block diagram, showing the condition of the western part of Indonesia in Devonian times.

subsidence is the result of a preceding cycle. Little is known however of the older orogenic cycles in Indonesia.

The arcuate Anambas-Schwaner-Telen zone turns its convex side towards the south. The orogenic cycle, which later on migrated to the south-west, away from this focus, is much more completely developed than the cycle which migrated in northerly direction. This may be interpreted as showing that the formation of the Anambas

130

geosyncline was the result of an older, pre-Devonian orogenic cycle, which was located in the region of the South China Sea.

The geosynclinal subsidence of the Anambas zone in Devonian times was apparently able to release sufficient internal energy to become, in its turn, a focus for two new cycles of orogenesis migrating away from it. One of them migrated towards the Gondwana foreland and the other towards the Asiatic hinterland.

Volcanic rocks of the ophiolitic suite intruded the Devonian sediments of the Anambas geosyncline. These have been indicated

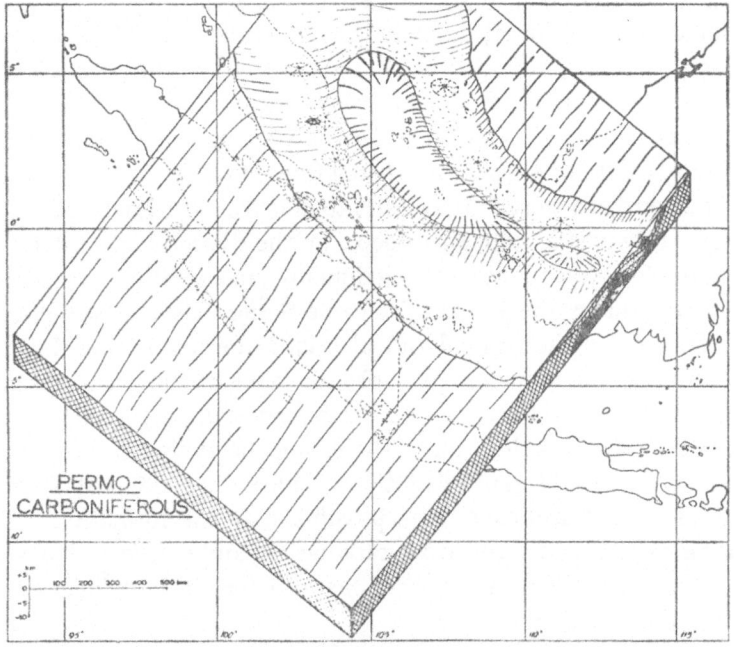

Fig. 35

Block diagram, showing the condition of the western part of Indonesia in Permo-Carboniferous times.

in black on the side of the block diagram, and the primeval continent has been indicated by cross hatching. The dots represent newly deposited Devonian sediments of neritic facies.

Permo-Carboniferous

Fig. 35 shows the next phase of evolution. A geanticline has been pushed up, as appears to be usual for such geosynclinal zones, and on both sides subsiding troughs have been formed.

The uplift of a geanticline is necessarily associated with the subsidence of the adjacent regions, as the total volume remains constant. In the depth, material is withdrawn from the adjacent areas and the overlying crust subsides. This volumetric compensation can take place on both sides of the geanticline as for example in the oldest orogenic phase of the Anambas cycle. If the uplift migrates in lateral direction, the material tends to come mainly from the convex side, and typical foredeeps originate.

The northern border deep, situated in the Natuna zone was a long narrow trough, in which the abyssal sediments of the Danau formation and the basaltic "Pulu Melaju" eruptive rocks were formed. In southerly direction the facies changes from abyssal to neritic and littoral, as we pass from a deep trough into a zone of islands. Along the coasts of these islands coral reefs were growing, or there were swampy coasts where peat was being formed.

The southern border deep, situated in the Karimata zone, was presumably wider and less deep; in the Permo-Carboniferous of eastern Malaya we find neritic sediments mixed with much volcanic material; this is the Pahang Volcanic Series. The sea also covered the Tin zone of the Sunda Land, at least in Permian times, but here the sediments do not show the volcanic facies of the Karimata zone.

According to de ROEVER, the Permian in Bangka consists of dynamometamorphic shales, sandy shales and fairly fine-grained quartz sandstones, with local intercalations of limestone beds with Permian index fossils (Fusulinas). It is possible that the formation, which contains detritus derived from radiolarites and siliceous shales and which was classified as Upper Triassic before, also belongs to the Permo-Carboniferous.

In this period the history of the Djambi nappes begins, the formation of which constitutes a difficult mechanical puzzle which we shall try to solve in this chapter.

In this zone a few thousands of metres of Permo-Carboniferous sediments of the Pahang Volcanic Series were deposited on a basement of pre-Carboniferous granites. Parts of this series are at present, that is two hundred million years later, found as nappes on the eastern flank of the Barisan Mountains, at a distance of 350–400 kilometres from the zone where these Permo-Carboniferous sediments were originally deposited.

The coarse conglomerates which are amongst others found in the

Salamuku beds of the Permo-Carboniferous series of the Djambi nappes, indicate that in the region of provenance at that time considerable contrasts in relief were in existence.

Permo-Triassic (Fig. 36)
After the Permo-Carboniferous period of sedimentation and volcanism in the two subsiding troughs of the Natuna and Karimata zones, these geosynclinal strips were raised to form mountains. The geosynclinal foredeeps consequently migrated outwards. In the

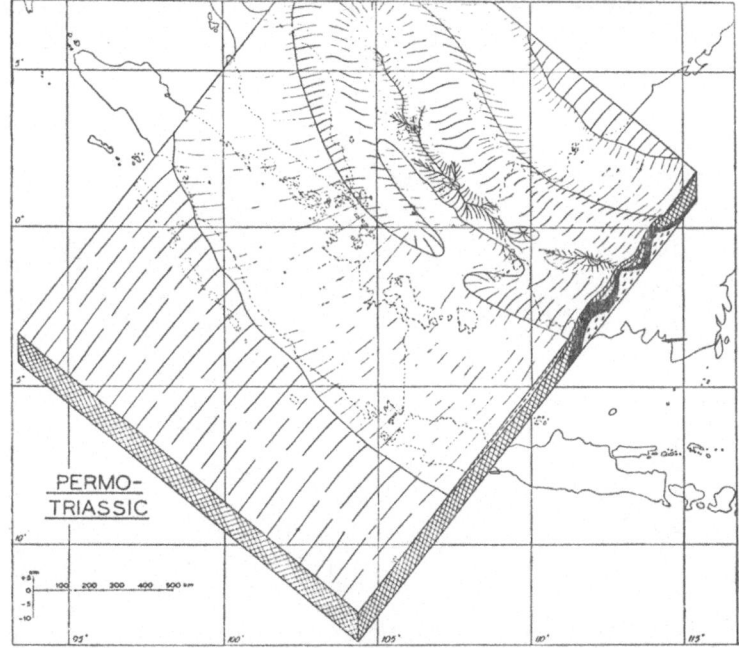

Fig. 36

Block diagram, showing the condition of the western part of Indonesia in Permo-Triassic times.

geanticlinal ridges granitic batholiths intruded; the tonalite masses of western Borneo belong to this phase of orogeny.

This uplift of the Natuna and Karimata zones and the associated folding of the Permo-Carboniferous strata was responsible for the formation of an angular unconformity between the Permo-Carboniferous and the overlying Upper Triassic in these zones and parts of the adjacent zones. Farther away from the zones of active

133

orogenesis sedimentation continued without interruption. This is shown, for example, by the succession of Batu Basi near Singkarak in middle Sumatra, where the Permian (consisting of sandstones, siliceous and marly shales, Fusulina limestones, radiolarites and tuff beds) is conformably overlaid by Trias (consisting of shales, marly limestones with fossils and tuff beds).

Late Triassic (Fig. 37)
After the Permo-Triassic orogenesis in the central structural zones of the Sunda region the relief decreased again. This was done partly

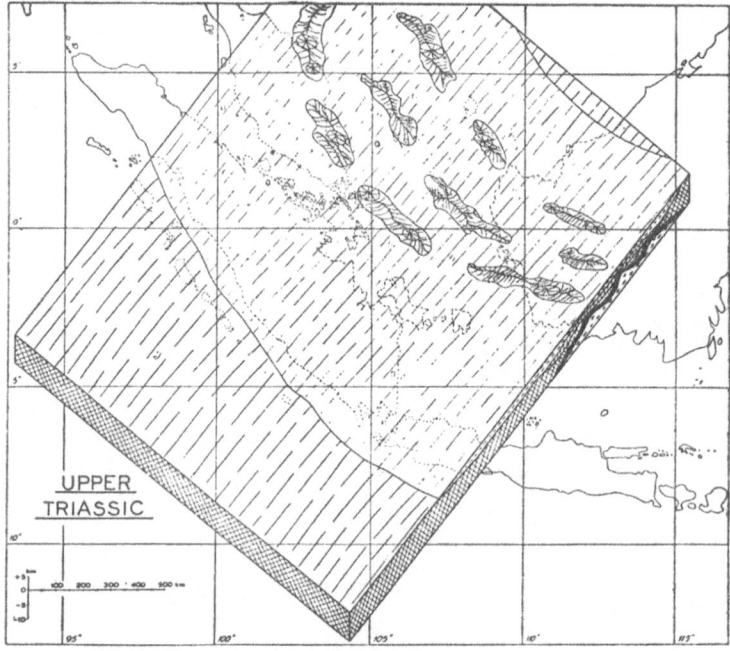

Fig. 37

Block diagram, showing the condition of the western part of Indonesia in Upper Triassic times.

by the denudation of the Permo-Triassic mountains, partly by the subsidence of the geanticlinal ridges. The result was that in Late Triassic times the sea occupied vast areas. In addition to this the mountain chains of the Natuna and Karimata zones became actively volcanic.

ZEYLMANS VAN EMMICHOVEN distinguishes in Western Borneo

134

Upper Triassic in a normal sedimentary and in a volcanic facies. The first consists of conglomerates, sandstones and shales. Limestones are lacking. The conglomerates contain detritus of granites, Permo-Carboniferous rocks and crystalline schists. The volcanic facies of the Upper Triassic consists of relatively alkali-rich eruptive rocks and tuffs (so-called keratophyres).

The Upper Triassic in the Riouw Archipelago (the Riouw formation) also consists of sandstones and shales with conglomeratic inter-calations which contain components of granites, radiolarites, quartz-ites, sandstones, shales, siliceous shales and minerals from crystal-line schists, as well as detritus of volcanic rocks. Limestone is rare in this formation. In Lingga and Bangka index fossils of Late Triassic age have been found in it. In the island of Valsch Durian, situated between Lingga and Kundur, the Riouw formation rests with a basal conglomerate on a formation of lustrous slates and amphibolites. It is impossible to decide whether the latter formation corresponds to the Permo-Carboniferous. The Riouw formation closes with a series of sandstones and shales, containing thin coal seams and leaf beds, in the island of Bintan. The age of these fossil plants is probably Late Triassic to Early Jurassic, presumably Rhaetic, according to JONGMANS.

According to de ROEVER the Upper Triassic of Bangka contains intercalations of sometimes porous basalts and andesites and their conglomeratic detritus. This indicates that these sediments were deposited in or near a zone of active volcanism after which they were transported to their present place by the folding, faulting and overthrusting of later date.

In general it is true that in Late Triassic times quiet development took place, mainly characterized by a slow epeirogenic subsidence of the crust and the denudation of the Permo-Triassic mountains. But in the depths, energy accumulated for the following orogenic evolution which took place towards the end of the Triassic or in Early Jurassic times. The active volcanism of the Permo-Triassic mountains in Late Triassic times can be considered as a forerunner of the important tectonic and magmatic events which characterise the next phase.

Early Jurassic (Fig. 38)
After the deposition of the Late Triassic sediments the Natuna and

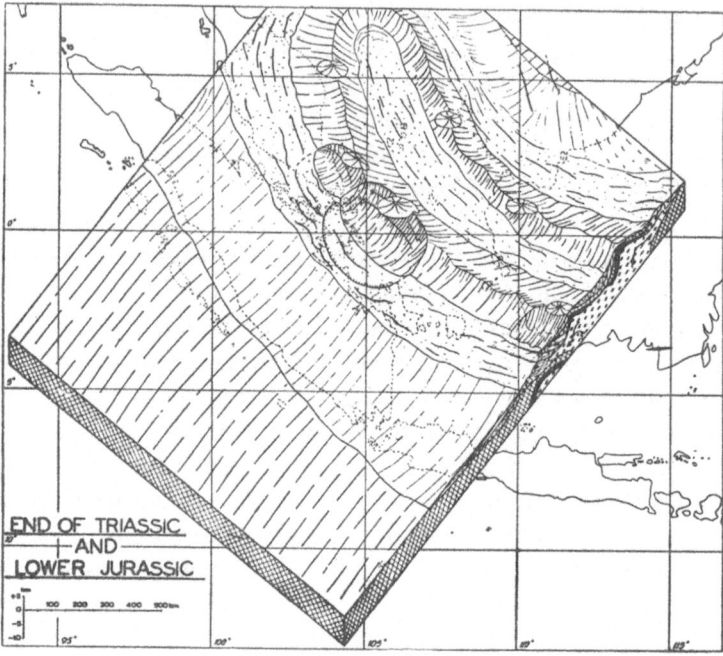

Fig. 38

Block diagram, showing the condition of the western part of Indonesia at the end of Triassic and in Lower Jurassic times.

Karimata zones were pushed up for the second time and gave rise to imposing, probably volcanic, mountain ranges. In the core of these geanticlines granite batholiths intruded for the second time. These granites are in general more homogeneous and richer in silica than the tonalitic granite intrusions of the preceding, Permo-Triassic orogenesis. The silica content is approximately $76\frac{1}{2}\%$.

The Late or post-Triassic movements raised the thick series of sediments which had been deposited in the previous periods. This sedimentary epidermis slipped like a cloak from the shoulders of the rising basement and piled up in the adjacent subsiding strips. But it was not only the sedimentary epidermis which was involved in this process of free gliding. The crystalline basement of these mountain chains had been severely broken by the Permo-Triassic orogenesis and showed many weak spots. In addition to this the basement had been melted and mobilized from below by the advancing granite masses and other magmatic processes. The result was that in several

136

places the mountains began to collapse, like an idol on feet of clay. Dermal gravitational tectogenesis therefore also took place. These processes may be considered as reactions to the arching up of the earth's crust. In the uplifted areas tensile stresses originated. The newly deposited, still unconsolidated sedimentary series began to flow and became thinner without showing clear signs of folding. In Bintan, for example, the Lower Mesozoic terrestrial formation shows dips of not more than 10°. In the adjacent subsiding zones strong folding occurred accompanied by overthrusting. Such a strongly folded strip runs for example across Western Batam (Bulan Straits).

During this orogenic evolution parts of the crystalline basement began to slide towards the foredeep in some sectors of the Karimata zone. These blocks carried on their backs series of unchanged Permo-Carboniferous and sometimes also Triassic sediments. As these blocks were gliding over plastic clays and sands of the Triassic formation they could cover large distances. Finally they ended up in the deepest parts of the foredeep, which at that time was situated in the Tin zone, that is more than 100 kilometres from their land of origin. The journey of the Djambi nappes therefore began at the end of Triassic or the beginning of Jurassic times with a first stage of over 100 kilometres.

Late Jurassic and Early Cretaceous (Fig. 39)
The descendant of the Anambas system which had been migrating in northerly direction, more or less petered out after the previous orogenic phase when it entered the older rigid parts of the earth's crust in the South China Sea. A shallow sea surrounded the remnants of the Triassic mountains there, for example in the Chinese Districts of western Borneo.

The cycle which migrated in a southerly direction away from the Anambas System, on the other hand, continued to develop. A new mountain chain was pushed up from the foredeep in the Tin zone. This was the Tin range, which was intruded by granite batholiths. These granites are the tin bearing granites of Malaya (Main Range), Singkep, Bangka and Billiton.

This Late Jurassic orogenesis gave rise to the Malaya Orogen, which consisted of two parallel chains. The younger outer arc of this orogen coincided with the Tin zone and was initially not yet volcanic. The inner arc was formed by the remnants of the older

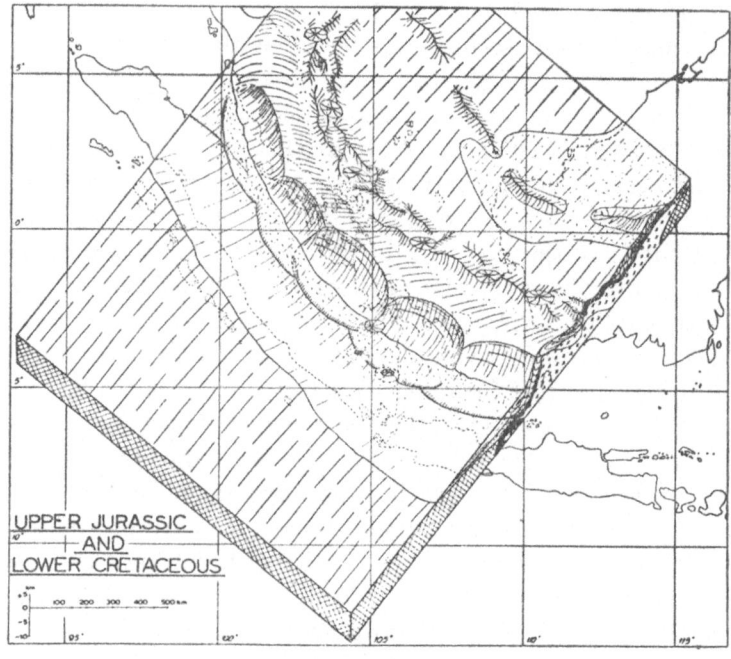

Fig. 39

Bock diagram, showing the condition of the western part of Indonesia in Upper Jurassic and Lower Cretaceous times.

volcanic chain in the Karimata zone. Subsequent erosion and the absence of fossiliferous sediments make it impossible to decide how far the denudation of the Karimata chain has progressed in Jurassic times, nor do we know whether this zone was lifted up again or whether external volcanism occurred.

The Tin range suffered the same fate as the Karimata chain had suffered before. The compressed sedimentary series in the Tin mountains could only form an unstable geanticlinal structure. This Tin range collapsed under its own weight and parts of it slipped towards the new foredeep which was situated where eastern Sumatra now is.

The granite wedges of the Djambi nappes with their cover of Permo-Carboniferous sediments began to glide towards the deeper part of this foredeep. During this second stage they covered again more than 100 kilometres. Parts of the Tin range with their intrusive Tin granites also began to glide towards the foredeep. This was the

138

first large scale gliding of the nappes containing tin ores. They are at present found on the eastern side of the Barisan (Suligi-Lipatkain nappe).

In Early Cretaceous times deposits were formed in this East Sumatra foredeep, which must have originated far from the land in a deep sea. These are the Lingsing beds, consisting of very fine radiolarian ooze with interstratifications of limestone beds and tuffs produced by eruptions of volcanoes at comparatively large distances. Ophiolitic magmas intruded the Lingsing beds.

More towards the east, at the foot of the Tin range, Lower Cretaceous deposits of an entirely different facies originated. They are the Saling beds, consisting of andesitic lava flows and breccias with intercalated reef limestones. These volcanic eruptions therefore must have been of the Pacific type.

These two entirely different types of Lower Cretaceous sediments which originally were deposited in widely separated areas, are at present closely folded and overthrust in the Gumai and Garba Mountains east of the Barisan of southern Sumatra. These over-thrusts are due to the later orogenic revolution in Middle Cretaceous times.

Middle Cretaceous (Fig. 40)
The subsiding trough in eastern Sumatra, which in early Cretaceous times formed the foredeep of the Tin range, was pushed up into a geanticlinal structure in Middle Cretaceous times. The axis of the foredeep consequently migrated farther to the south-west, to the zone which is at present occupied by the Barisan of Sumatra and the geanticline of southern Java.

The East Sumatra range suffered the same fate as its predecessors to the east. The mountain chain of eastern Sumatra collapsed and the fragments formed a pile of nappes in the foredeep, situated in the Barisan zone. The deepest tectonic unit of this zone is formed by the intensely folded, autochthonous sediments, the so called "Old Slates". They are successively overlaid by the nappes of the Lower Cretaceous (coming from eastern Sumatra), the unit coming from the Tin zone (Suligi-Lipatkain nappe) and the units coming from the Karimata zone (Djambi nappes).

For the Lower Cretaceous nappes this was the first time they had been gliding; they covered a distance of about 100 kilometres. For het nappes from the Tin zone this was the second phase of gravitational tectogenesis; it raised the total distance covered to more than 200 kilometres. For the Djambi nappes this was the third stage which left them at approximately 350 kilometres from their area of provenance.

Due to these events the youngest sediments (those of the foredeep) are now found at the bottom of the pile. Consequently they have

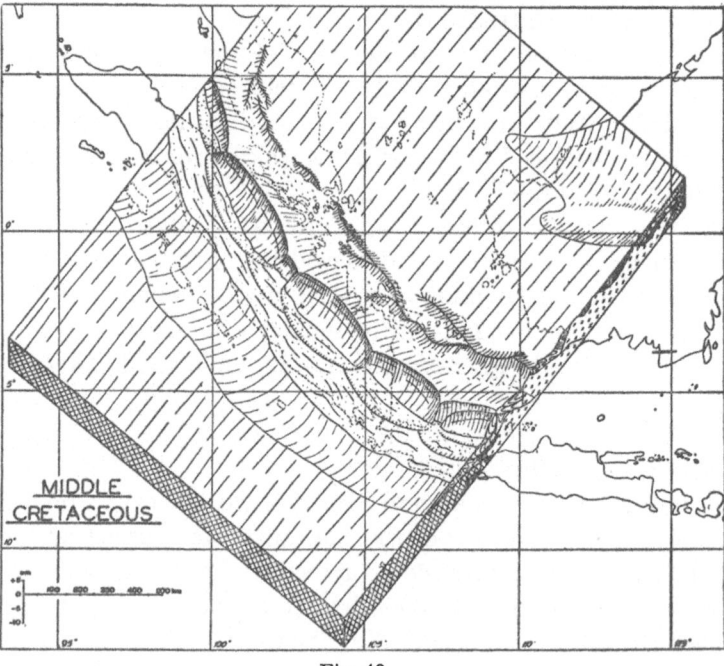

Fig. 40

Block diagram, showing the condition of the western part of Indonesia in Middle Cretaceous times.

suffered the strongest changes in texture and mineralogical composition. We may compare these "Old Slates" of the Barisan zone with the "Schistes lustrés" or "Bündner Schiefer" of the Alps in western Europe. The oldest sediments, i.e. the Permo-Carboniferous of the Djambi nappes, on the other hand, are now near the top of the tectonic structure. Despite the fact that they covered the largest distance (three times they covered a distance of more than 100 kilometres) these Permo-Carboniferous sediments have suffered less change than the rest. This is so because they have never been

140

covered by higher tectonic units. They contain well preserved fossils (Fusulinas) and a flora of Late Carboniferous age, which has been described by JONGMANS and GOTHAN.

The position of the Djambi nappes in this pile of overthrust masses may be likened to that of the East Alpine nappes in the European Alps which also form the highest units and have covered the largest distance. In Fig. 42 the three phases of collapse are

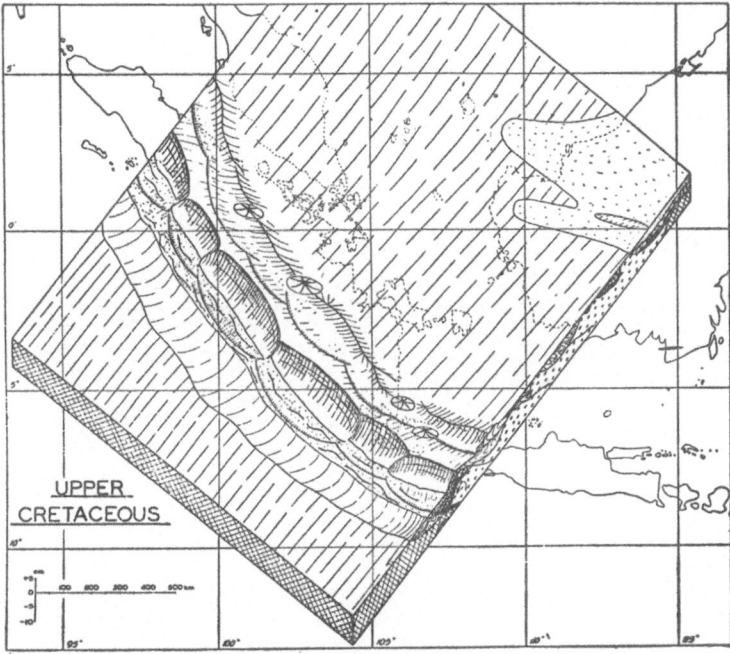

Fig. 41

Block diagram, showing the condition of the western part of Indonesia in Upper Cretaceous times.

shown, which are responsible for the tectonic structure of the Barisan zone.

Late Cretaceous (Fig. 41)

In Late Cretaceous times the geosynclinal trough of the Barisan zone was for the first time transformed into a mountain chain and the axis of the foredeep migrated another 200 kilometres to the southwest. At that time the foredeep reached the zone where at present the outer arc of the Sunda Mountain System is situated (Nias-Enggano zone).

141

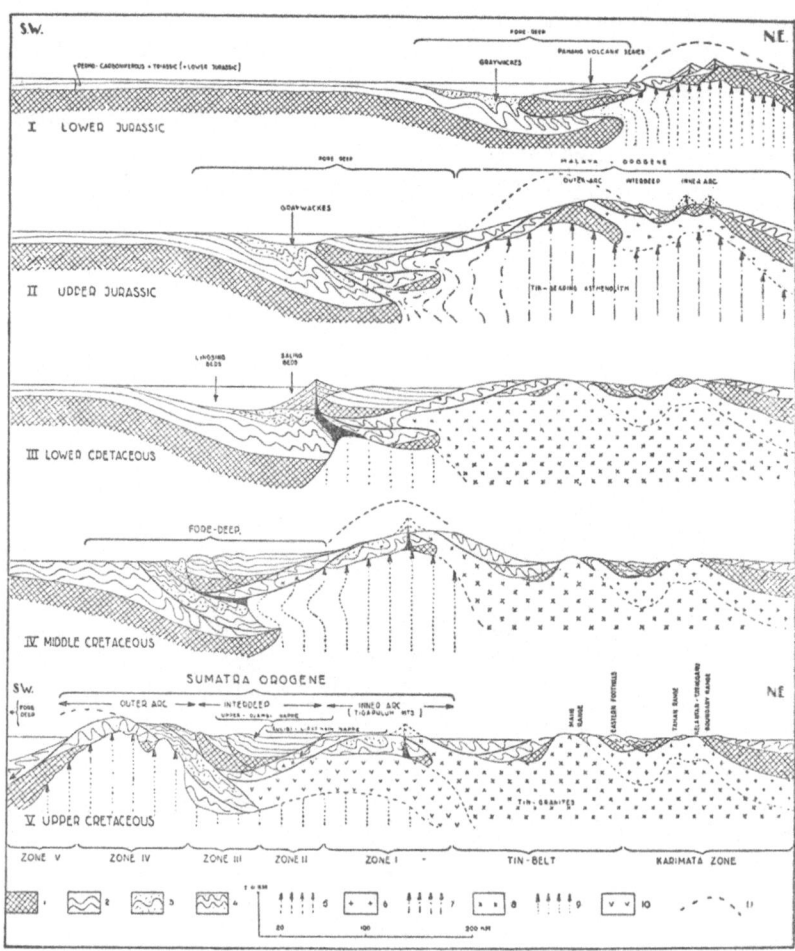

Fig. 42

Five schematical sections, illustrating the pre-tertiary evolution of the Malay
Peninsula and Sumatra.

1. Pre-permo-carboniferous basement complex. 2. Sedimentary epidermis of young
palaeozoic and mesozoic age, without further distinction of age. 3. Graywacke
formation (Jurassic?). 4. Lower Cretaceous Saling beds. 5. Asthenolithic migma
and magma level of the Karimata zone in the Lower Jurassic. 6. Idem consolidated
(crystallized) in the Upper Jurassic. 7. Asthenolithic migma and magma zone of
the Tin Belt in the Upper Jurassic. 8. Idem consolidated (crystallized) in the Lower
Cretaceous. 9. Asthenolithic migma and magma level of the Sumatra Orogene in the
Middle and Upper Cretaceous. 10. Idem consolidated (crystallized) in the Upper
Cretaceous. 11. Outlines which the geanticlinal uplifts from the foredeeps would
have reached if no bathydermal spreading had occurred.

142

The Primeval or Proto Barisan was also not able to withstand the force of gravity; it collapsed and fragments began to glide towards the foredeep. Sediments and wedges of the crystalline basement were pushed together in the foredeep and reversed faults and overthrusts originated. Locally this compression in the foredeep caused the rocks to emerge above sea level and consequently angular unconformities developed between the pre-Tertiary and Tertiary sediments. The fact that the rocks in the foredeep were locally forced up above sea level, does not mean that primary tectogenesis had already started. Primary tectogenesis in the Barisan foredeep did not start before the end of Tertiary times. On the contrary, towards the end of Cretaceous times a stronger subsidence of the basement began in the Nias-Enggano zone. This was due to the volumetric compensation of the rising Barisan range. During the subsidence of the Nias-Enggano zone ophiolitic magma intruded the strata in this foredeep.

The nappe structures which had been piled up in the Barisan zone during the previous orogenesis, found themselves on the eastern flank of the Proto Barisan when this geanticline was pushed up. They had, as it were, been overtaken by the geanticlinal crustal wave and could no longer move towards the new foredeep. Instead, there is rather a tendency to glide towards the backdeep, a kind of backward folding, similar to that known from the nappes in the Alps during the last, Insubric phase.

Granite batholiths intruded the core of the Proto Barisan during and immediately after the Late Cretaceous orogenesis, but external volcanism did not yet occur.

Little is known about the fate of the older chains in eastern Sumatra, as the greater part of this area is concealed from view by Tertiary sediments. It is possible that in Late Cretaceous times more periods of uplift took place, accompanied by active volcanism. In the Tigapulu Mountains plutonites and volcanites of presumably Late Mesozoic age occur.

At the end of Cretaceous times, the Sumatra Orogen had been formed, consisting of a non-volcanic young outer arc (the Proto Barisan) and an extinct volcanic older inner arc. This Sumatra Orogene lies approximately 400 kilometres to the south west of its predecessor, the Malaya Orogen, which existed 60–70 million years earlier.

The migration of the orogenesis which started from the Anambas zone had, however, not yet come to a stop, as will be shown by the events in Cainozoic times. Much more data are available for this Cainozoic history and we shall therefore deal with it at some greater length.

Eocene (Fig. 43)

Before we proceed with the description of the orogenic evolution of the Sunda region, we must first pay some attention to the epeiro-

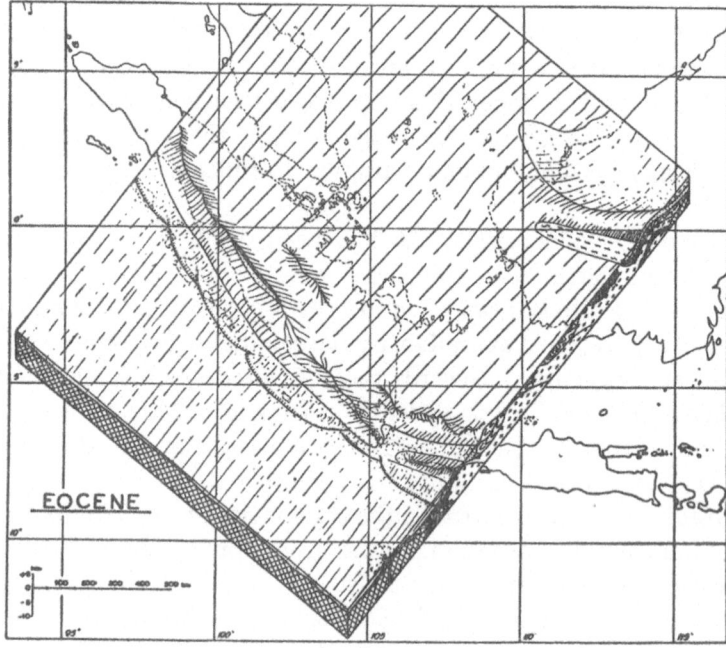

Fig. 43

Block diagram, showing the condition of the western part of Indonesia in Eocene times.

genic movements which occurred in the neighbourhood of this region.

In the preceding block diagrams, the region of the Indian Ocean to the south west of the Sunda region was shown as land. A border-land was presumably situated there, a remnant of the Palaeozoic Gondwana land, which has been called Olim-Gondwana by STILLE.

Towards the end of the Palaeozoic and in the course of the

144

Mesozoic Eras the breaking up and subsidence of that part of the earth which is now occupied by the Indian Ocean took place. This slow epeirogenic subsidence is presumably due to slow chemical transformations which changed the density of the various silicate shells in the tectosphere. These physico-chemical processes in the depths were followed by slow epeirogenic earth movements over vast areas. These movements are the surface expression of mass displacements striving after restoration of the hydrostatic equilibrium in the silicate shell.

There are still many geologists and geophysicists who question the possibility of the subsidence of continental areas to oceanic depths, as this violates the principle of isostasy (floating equilibrium). The observations, however, must take precedence of all theories. Recent oceanographic investigations, such as those of EWING and his collaborators in the western part of the North Atlantic Ocean, show that the subsidence of vast areas of the crust has in fact taken place. The same has recently been maintained by FAIRBRIDGE[1] for the Indian Ocean.

The possibility therefore exists that the principle of isostasy, which postulates that the various densities remain constant, may have to be replaced by a hypothesis which admits that chemical changes in the depths can change the density to such an extent that important epeirogenic reactions result.

The Christmas Island volcano began its activity on the boundary of the Mesozoic and Cainozoic Eras. The products of this volcano belong to the Atlantic suite which elsewhere occurs in disintegrated and partly foundered continental areas. The activity of the Christmas Island volcano may therefore be taken to indicate that a similar process of destruction of Olim-Gondwana began at that time. Moreover, the deep earthquakes which at present occur in the Sunda region indicate that the adjacent part of the ocean floor is still subsiding relative to south eastern Asia.

The following tectonic events of local importance occurred during Eocene times. Due to tensile stresses an elongate graben originated in the culmination of the Proto Barisan. In this graben coarse conglomerates were deposited in some places, as for example the Brani conglomerate near Mangani in middle Sumatra. Elsewhere,

[1] "The juvenility of the Indian Ocean". Scope, Journal of Science Union, University of West Australia, 1948, Vol. 1,3, p. 29–35.

fresh-water lakes formed, as for example in the basin of the Umbilin river in middle Sumatra, where marly sediments were deposited, containing beautifully preserved fossil fresh-water fishes.

Apart from this, the Eocene was a period of denudation of the Proto Barisan which during this period was not yet volcanic. The axis of this non-volcanic outer arc ran from Sumatra towards the east, along what is at present the south coast of Java. To the north of this eastern extension, for example in the Bajah region of Bantam (Western Java) sandstones and clays alternating with coal seams were deposited. KOOLHOVEN showed that the sediments were derived from a non-volcanic ridge, situated to the south where at present the Wijnkoop Bay is found (now more than 3,000 metres deep).

In middle Borneo orogenesis also continued. The Semitau ridge, which forms the extension of the Early Mesozoic Natuna zone, showed some vertical movements in Cretaceous times. Between this barrier and the Anambas-Schwaner Mountains zone the Seberuwang Basin was formed, where a series of Cretaceous sediments of more than 3,000 metres thickness were deposited. Towards the end of Cretaceous times this series was folded and the Semitau barrier pushed up into a non-volcanic mountain ridge. This was followed in Eocene times by a general subsidence and the Seberuwang region became a brackish water basin of about 350 kilometres length and 100 kilometres width. In this basin the Melawi beds were deposited which attain enormous thicknesses (4500 metres to 9700 metres according to BELTZ). The area to the north of the Semitau barrier was open sea in Cretaceous and Eocene times; thick marine deposits originated here.

Oligocene (Fig. 44)
In Oligocene times Olim-Gondwana continued to subside. This subsidence was accompanied by the foundering of the border zone of the Sunda Land, causing the Proto Barisan to disappear slowly below sea level. Finally only the highest tops were left above sea level and formed islands. The sea transgressed over eastern Sumatra, preceded by the deposition of sand, gravel and peat layers in the drowning river valleys. This is when the Quartz Sandstone Formation, which comprises the coal seams of Umbilin, originates.

At several points, but particularly in its southern part, the Proto Barisan became actively volcanic. The same happened in southern

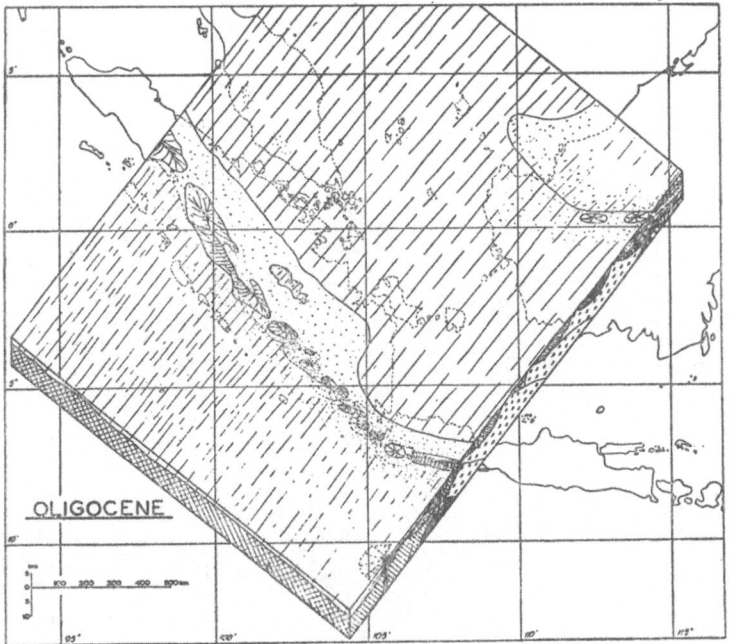

OLIGOCENE

Fig. 44

Block diagram, showing the condition of the western part of Indonesia in Oligocene times.

Bantam. This volcanism is partly submarine and the "Old Andesite formation" consequently contains intercalations of Oligo-Miocene limestones and tuffaceous marls.

In Oligocene times the Semitau zone in Borneo also became actively volcanic (Müller Mountains). Its development is therefore similar to that of the Barisan. A volcanic stage of development succeeded the non-volcanic one.

In general it is true that the Eocene and Oligocene were periods of comparatively quiet development. This evolution continued into Middle Miocene times.

(Middle) Miocene (Fig. 45)

In Middle Miocene times the geanticline of the Barisan and southern Java was for the second time pushed up. This time the entire chain was volcanic. The first volcanic cycle which produced the predominantly basaltic and andesitic eruptions of the Old Andesite for-

147

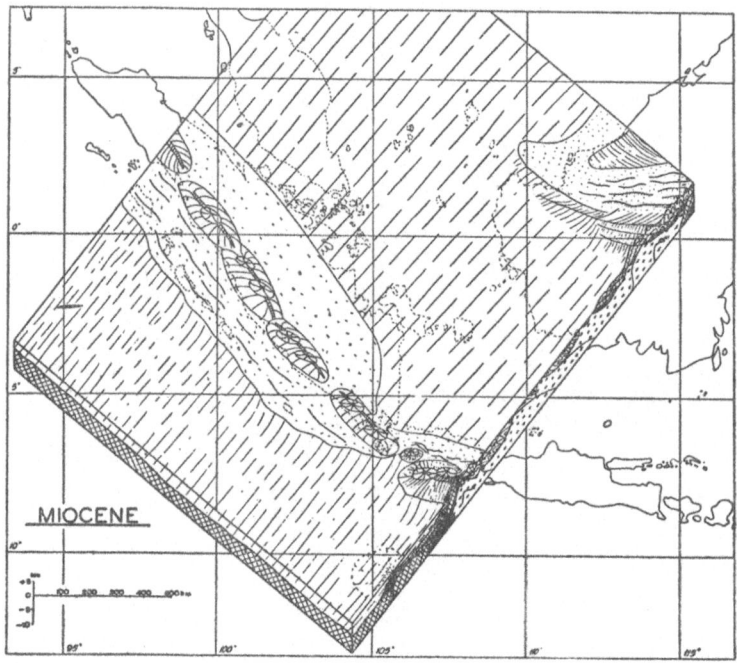

Fig. 45

Block diagram, showing the condition of the western part of Indonesia in Miocene times.

mation, was terminated by violent eruptions of tuffs of a more acid (granodioritic and granitic) composition. The latter material erupted through fissures in the culmination of the geanticline.

Granodiorites and granites intruded the core of the Barisan-southern Java geanticline during and after this Intra-Miocene uplift. These batholiths of acid plutonic rock were accompanied by dacitic and liparitic intrusions and they soaked the country rock with hot metalliferous solutions. Thus originated the gold and silver ores of Redjang Lebong, Tambang Sawah, Simau, Tjikotok, Tjipitung, the Djampangs, etc.

The arching up of the Barisan-southern Java geanticline therefore caused tension, accompanied by volcanism and the formation of ore deposits in the culmination. In the foredeep, on the other hand, which was situated where now the outer arc is, folding and upthrusts in south-westerly direction took place at the same time in the Early to Middle Tertiary sediments. The tension in the uplifted

148

strip and the compression in the adjacent zone of subsidence probably compensate each other as discussed in the chapter on gravitational tectogenesis.

It is remarkable that the sediments in the subsiding trough of eastern Sumatra, on the other side of the Barisan ridge, have hardly or not at all been disturbed by this Intra-Miocene orogenesis. Here we find a conformable succession of Telisa beds and Lower, Middle and Upper Palembang beds; this means that sedimentation continued from Oligo-Miocene into Quaternary times. In Telisa times the sea transgressed towards the east; later, in Palembang times it regressed towards the west (cf. Fig. 8). During the Middle Miocene uplift of the Barisan the situation in the sedimentary basin of eastern Sumatra was apparently not yet ripe for the occurrence of gravitational tectonic reactions.

In the outer arc (i.e. the Nias-Enggano zone), on the other hand, the older Miocene is unconformably overlaid by the marine Kalea formation, which in age approximately corresponds to the Middle Palembang beds. This means that here folding has taken place in Late Miocene times.

In the Java sector the situation is slightly different. Nothing is known about Middle Miocene compression phenomena in the outer arc south of Java, as the deposits have not yet been lifted above sea level. In Java itself, however, it appears that the geanticline in southern Java collapsed and that parts of it moved towards the basin of northern Java, — this in contrast to Sumatra. In northern Java we find everywhere a clear angular unconformity between Lower and Upper Neogene.

In central Borneo a great orogenic revolution had taken place in Miocene (or somewhat earlier, in Oligo-Miocene) times. The Semitau zone, which in Cretaceous and early Tertiary times had already been pushed up several times, was energetically lifted up towards the end of Palaeogene times. Due to this primary-tectonic uplift the Semitau zone collapsed and parts of it moved towards the foredeep situated in the north. In this foredeep, where thick sedimentary series varying in age from Permo-Carboniferous to Eocene had been deposited, a pile of nappes was formed.

The Eocene deposits in the foredeep of the Semitau zone in central Borneo found themselves at the bottom of this pile. The same thing had happened to the autochthonous "Old Slates" of the Barisan, which now belong to the deepest structural units of this mountain chain. The Eocene sediments of central Borneo were also metamorphosed into lustrous slates. In the Pennine nappes of the western Alps similar Eocene slates occur, which also form the lowest tectonic units. These Pennine nappes originated during the Middle Tertiary main phase of Alpine orogenesis; they are therefore of the same age as the nappes with Eocene lustrous slates in central Borneo. The two areas are also comparable in the intensity of the movements.

In connection with these Middle Tertiary tectonic processes in central Borneo the following remark must be made. In the chapter on geophysics the crustal buckling hypothesis of VENING MEINESZ was mentioned as an explanation of the negative isostatic anomalies in the outer arc. The buckling, in this view, occurred during the Middle Tertiary main phase of folding and overthrusting in the outer arc of the Sunda Mountain System, which zone coincides with the VENING MEINESZ zone. This hypothesis assumes that both the buckling of the granitic crust and the folding and overthrusting of the sedimentary epidermis are the result of a horizontal (tangential) compressive force in the crust, coming from the N.N.W.

When in the early thirties this hypothesis was framed, little was known about the geology of central Borneo. This area was left blank on the maps of UMBGROVE and VENING MEINESZ. Since then the investigations and publications of ZEYLMANS VAN EMMICHOVEN have provided much information about this area. It appears that during the Tertiary main phase of folding and overthrusting in the outer arc of the Sunda Mountain System, intensive alpine-type earth movements took place in central Borneo. It is therefore doubtful whether the central part of the Sunda region, i.e. central Borneo would have been able to transmit this tangential compressive force from the Asiatic continent to the outer arc of the Sunda Mountain System.

Pliocene (Fig. 46)
After the orogenic revolution in Miocene times, a period of quiet development followed, especially in the southern part of the Sunda region. The relief contrasts in Sumatra and Java decreased and the sea advanced again over the areas which it had temporarily left.

This period of apparent rest was however the incubation period

150

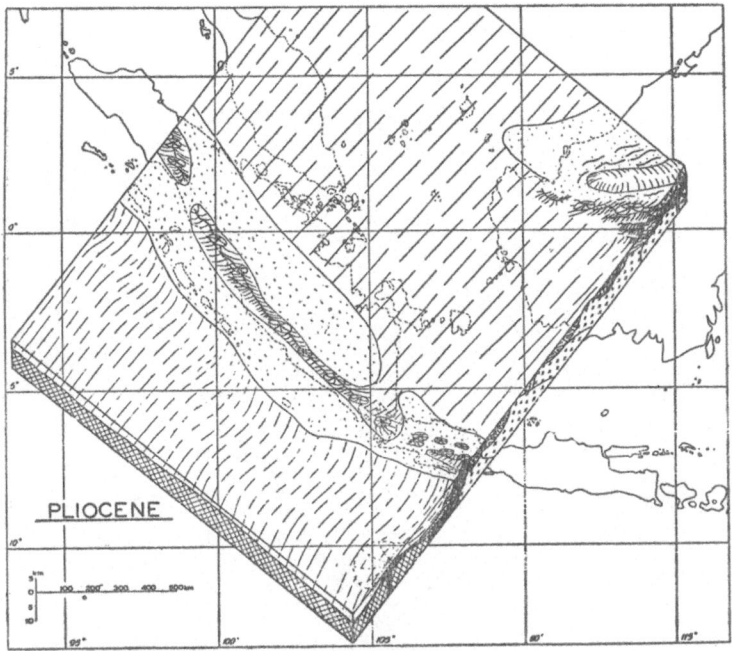

Fig. 46

Block diagram, showing the condition of the western part of Indonesia in Pliocene times.

of the next orogenic revolution. In Pliocene times energy accumulated in the depths for another uplift.

To the north, in Borneo, a mountain chain rose from the compressed foredeep and formed the Upper Kapuas Mountains. The axis of the foredeep consequently moved farther outwards to Serawak in north-western Borneo.

Pleistocene (Fig. 47)

The Pleistocene in Indonesia was a period of powerful mountain building. In some areas the orogenic movements started in Pliocene times and in many parts they continued into Holocene times. This youngest orogenic phase occupied approximately one million years. This is very long if measured with our human yardstick of time, but comparatively short compared with the length of the periods of more quiet evolution with which these tectonic revolutions alternate.

A more detailed analysis of the time and place of the tectonic

151

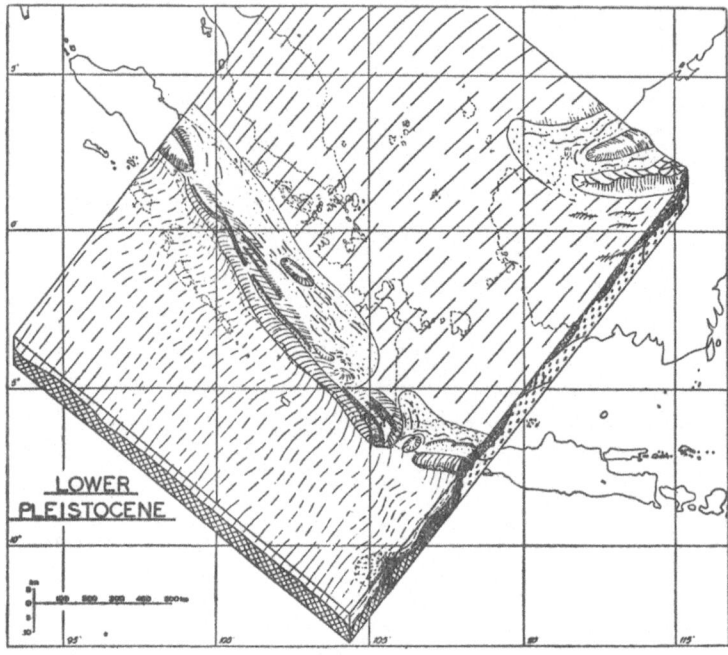

Fig. 47

Block diagram, showing the condition of the western part of Indonesia in Lower
Pleistocene times.

movements shows that the youngest orogenic phase as a whole
comprises approximately one million years and even now has not
yet come to an end in some areas, but also that in different sectors
the tectonic revolution started at different times. Such an analysis is
possible for Java, thanks to the detailed Quaternary stratigraphy
based on vertebrates. In the geosynclinal basin of northern central
Java orogenesis started in Late Pliocene times, but in easterly and
westerly directions the tectonic movements started at successively
later dates. In the Kendeng zone of eastern Java, for example, we
find that in the most western sector of the anticlinorium folding started in early Pleistocene times, that in the central part it started in
Middle Pleistocene and in the eastern part, where the axes of the
folds plunge under the Brantas delta, in Late Pleistocene times.
In the Madura Straits, which form the eastern extension of the
Kendeng zone, geosynclinal subsidence is still taking place and
tectogenesis has not yet started. The moment when tectogenesis

152

began therefore shifts through a period of one half to one million years if we move along the axis of an orogenic zone over a distance of only approximately 400 kilometres. In northern central Java this orogenesis has since the end of the Pliocene given rise to an important mountain chain: the North Seraju Mountains. In the Kendeng zone the uplift of the central sedimentary basin did not start before Quaternary times, and in the Madura Straits the uplift has not yet begun. This shift in the evolutionary phases of the various sectors of the orogenic zone of northern Java is also expressed in the isostatic anomalies. The North Seraju Mountains do not show a deficiency in gravity according to the theory of floating equilibrium. The uplift of these mountains seems to be complete. The Kendeng zone and the Madura Straits, however, show negative isostatic anomalies of more than 50 milligals, which indicates that in the future a considerable uplift will occur before the earth's crust is again in floating equilibrium.

This example is a warning for those geologists, who tend to consider orogenic phases as events occurring over vast areas or even over the whole earth, and at the same time. An orogenic phase, however important for a given area, has only local significance. It is neither a regional, nor a world-wide punctuation mark in the evolution of the earth. There is no direct, causal connection between the youngest orogenic phase in California, the so-called Pasadenic phase of STILLE, and the youngest orogenic phase in Indonesia which we are at present discussing.

It is, however, reasonable to assume the following. If, from a given focus of diastrophism, the orogenic process migrates step by step in successively wider arcs in a lateral direction, each of these arcs will be in a certain phase of tectonic evolution. The example of the Kendeng ridge in Java shows that within the same arc relatively great differences in the time and intensity of the orogenic phase can occur, but that for the rest such a mountain chain or island arc can be considered as a structural unit.

It appears also that the foci of diastrophism are arranged in belts which stretch across vast areas. The above mentioned Anambas centre for example belongs to a series of foci of diastrophism, which are situated in the wide Tethys sea. Starting from such rows of foci the orogenic systems migrate laterally as wave-like deformations of the crust. In each phase of development there are certain similarities between these geanticlines. Together they form long mountain chains, such as the Alpine system in Europe, the Himalayan system in central Asia and the Sunda system in south-east Asia and Indonesia.

The Sunda Mountain System consists, from north-west to south-east, of the Arakan Yoma arc around the Shan focus, the Andaman and Nicobar island arcs around the Mergui focus, the Sumatra-Java arc around the Anambas focus, the Lesser Sunda Islands around the Flores focus and the Banda arc around the Banda focus.

It will be clear that in this way mountain systems of striking uniformity can originate, despite the local differences in the nature and intensity of the orogenic evolution. The Sunda Mountain System, which is more than 7,000 kilometres long, forms a festoon consisting of about five arcuate sectors. In each sector we can distinguish the following elements: an active or extinct volcanic inner arc, a non-volcanic outer arc which is still incomplete and shows negative isostatic anomalies, and a foredeep.

We have only dealt with the evolution of the Sumatra-Java sector but a similar sequence of events took place — mutatis mutandis — in the other sectors of the Sunda Mountain System.

The above description will have shown how it is possible that large scale geotectonic features can originate, despite the fact that an orogenic phase is only of local significance.

The deficiency of gravity in the Vening Meinesz zone is a geophysical phenomenon which only originated during the last orogenic phase of the Sunda Mountain System. This zone coincides with the non-volcanic outer arc of the present mountain system. From the above description of the origin of these mountain chains it follows that in earlier evolutionary phases the more inward structural zones must have been underlaid by an uncompensated mountain root. In other words, each orogenic revolution must be accompanied by temporary zones of isostatic anomalies. During the older phases these zones were situated closer to the focus of diastrophism; the geophysical phenomena such as gravity anomalies and earthquakes move outwards from the focus of diastrophism in the same way as the orogenic deformation of the crust.

Let us now return to the discussion of the youngest orogenic phase in the Sunda region.

In Pleistocene times the Java-Sumatra arc was pushed up for the third time. The system of normal faults and grabens on the culmi-

154

nation of the Barisan (the Semangko zone) acquired its present structure during this phase. Eruptions of acid, pumice bearing tuffs are associated with these young fissures, as was the case during the Intra-Miocene phase.

Ranau in the southern and Toba in the northern part of Sumatra belong to this category of Pleistocene eruptions of granitic or granodioritic magma. In the chapter on volcanology we pointed out that there is presumably a causal relation between the arching up of the Batak tumor (300 kilometres long, 150 kilometres wide and more than 2000 metres high) and the explosive eruption of the top part of a granitic magma intrusion. This granite batholith came very close to the surface and found a way out through fissures near the top. It seems therefore natural to assume that the arching up of the Batak tumor is due to the pressure of magma which was pushed up from the depths.

As the Batak tumor is only a culmination in a much larger structure (the Barisan geanticline), it also seems natural to assume that the uplift of the entire Barisan and other similar geanticlines is due to magma pushing from below.

At the southern extremity of the Barisan another culmination was formed. This is the Sunda Straits tumor which in Pleistocene times also produced enormous quantities of acid tuffs, such as the Lampong tuffs in southern Sumatra and the Bantam tuffs in western Java. In the chapter on tectonics we discussed the collapse of this tumor, which took place in Late Pleistocene times, as an example of bathydermal gravitational tectogenesis. The Bajah tumor in southern Bantam passed through a similar development, but on a smaller scale.

The arching up and subsequent collapse of the Barisan of Sumatra and of the geanticline of southern Java was accompanied by the compression of the sedimentary series in the adjacent basins of eastern Sumatra and northern Java. In Java the collapse of the geanticline was compensated by upthrusts in a northerly direction in the geosynclinal zone of northern Java.

It is impossible to say how far the Plio-Pleistocene uplift of the Sumatra-Java geanticline caused gravitational collapse also in southerly direction, as the interdeep is covered by the ocean. Its probable occurrence in the Sunda Straits has already been mentioned at the end of Chapter IV. The breaking up of the Barisan Mountains

155

into a great number of parallel horsts and grabens with strong seismicity, indicates that parts of these mountains are still moving towards the ocean.

During the Liwa earthquake in 1933 open fissures, parallel to the axis of Sumatra, originated. The negative isostatic anomalies in the Semangko zone near Toba and Padang (shown on DE BRUYN's new isogam map, published in 1951) point to a local mass deficiency. In the Toba region this can be partly explained as due to the explosive removal of part of the batholithic magma chamber; but the collapse of this graben system caused by the lateral movement of fault blocks may also have something to do with these negative gravity values. No parts of the Barisan are at present moving towards the back deep, which has few earthquakes. But in the flank which disappears beneath the Indian Ocean, and in the adjacent basin which separates the Barisan from the outer arc, many tectonic earthquakes occur. These are presumably due to compression which compensates the extension of the Barisan. Under our eyes, but imperceptibly slowly, a geanticline is collapsing.

The collapse of the geanticlinal structure in Java took place for the greater part in Middle Pleistocene times, when the ape-man Pithecanthropus erectus was living there. The facies of the Kabuh beds (cross bedded sands and gravel which contain its fossil remnants) testify to the ever changing environment in which it was living.

The uplift of the outer arc began in the Pleistocene and still continues at present, in the Holocene. Before that the outer arc had been a subsiding zone for several tens of millions of years. In Cretaceous and especially in Tertiary times a considerable subsidence of the basement had taken place when this zone formed the foredeep of the Sumatra-Java chain.

During the successive uplifts of the Sumatra-Java geanticline the sediments in this foredeep were compressed. Locally and temporarily this forced the rocks above sea level, giving rise to angular unconformities in the sedimentary series. These movements, however, were not due to independent uplifts of the basement under this geosynclinal strip. The uplift of this structural zone by its own mountain root did not begin until Quaternary times; the negative isostatic anomalies indicate that the outer arc has not yet reached floating equilibrium.

In the northern part of the Sunda region, in Borneo, powerful orogenic movements were also taking place in Quaternary times. The Upper Kapuas Mountains, which had already been pushed up in Pliocene times, became volcanic; this is proved by the presence of Early-Quaternary volcanic material in the Nieuwenhuis Mountains and the Apo Kajan. This volcanism, at present, has become extinct.

In Serawak and Brunei a new mountain ridge, the Ularbulu ridge, was pushed up from the foredeep of the Upper Kapuas Mountains in Pleistocene times. The Late Neogene sediments around this uplift were consequently folded.

In the north-western coastal strip of Borneo, where the oilfields of Miri and Seria are situated, two periods of folding can be distinguished. The older one occurs on the boundary of Miocene and Pliocene times and is presumably connected with the movements of sedimentary strata, gliding from the then high area of the South China Sea towards the south east (Miri anticline). After these movements the South China Sea block began to subside. This subsidence, together with the uplift of the Ularbulu ridge, created stresses in the opposite direction. The younger, post-Pliocene folding was due to this new set of stresses (Seria to the north east of Miri).

This Plio-Pleistocene orogenesis created in central and north western Borneo a mountain system which — in contrast to most others — turned its concave side outwards, to the north west.

The inner arc of this Serawak Orogen is formed by the extinct volcanic Upper Kapuas Mountains, and the outer arc by the non-volcanic Ularbulu ridge.

The present state of the Sunda region (Fig. 48)
The Pleistocene tectogenesis is for the greater part responsible for the conditions which we observe at present. The main changes in the distribution of land and sea which occurred after this tectogenesis were due to the melting of the Pleistocene inland ice at higher latitudes, causing a world-wide rise in sea level of 80–100 metres. The old Sunda Land, which erosion had worn down to a peneplain, became submerged after this rise of sea level and the extensive Sunda Shelf originated. The floor of this shallow sea still shows the valleys of the rivers which drained this old land area (cf. Fig. 3).

During the Holocene the coast line again advanced quickly by accretion. In Sumatra this advance of the coastline took place at a

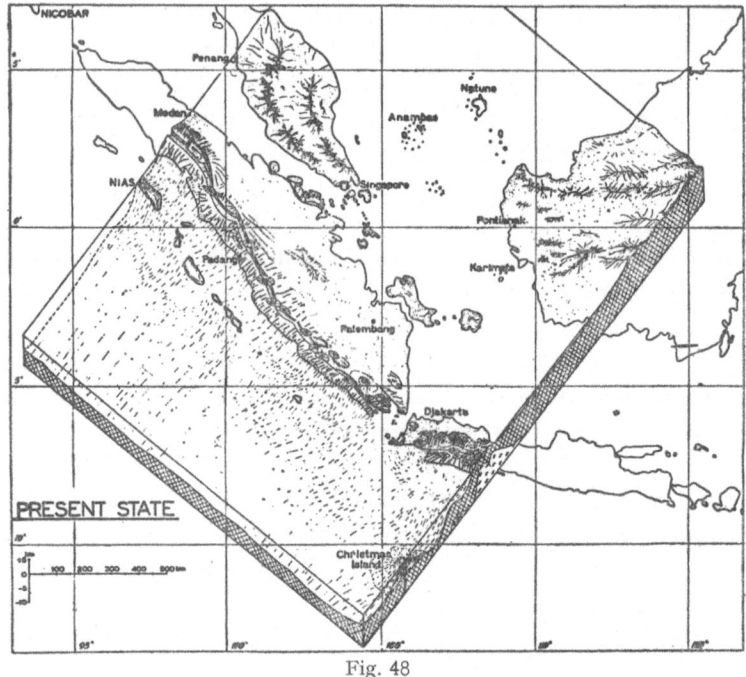

Fig. 48

Block diagram, showing the present condition of the western part of Indonesia.

rate of 100 metres per year. This means that at the beginning of the Christian era several bays reached as far inland as the foothills of the Barisan.

Near Semarang on the north coast of Java accretion takes place at a rate of 10 metres per year. In the Middle Age the sea reached the base of the Tjandi hills.

In the Riouw Archipelago, however, the coastline is retreating. When MARCO POLO was travelling from Indo China to Sumatra in 1292 he had to sail round Billiton. At that time the routes of commerce from India to China were controlled by the Hindu states in southern Sumatra. It is only later that erosion made the Singapore Strait navigable, a welcome short cut in this route. The result was that after 1350 Singapore and Malaya became more and more powerful. In the same period the sea power of southern Sumatra was increasingly hampered by the silting up of the bays on the east coast. This led finally to the total political decline of southern Sumatra.

The collapse of the Sunda Straits tumor was also not completed

158

until late Quaternary times. The Strait between Sumatra and Java has become navigable probably only in historic times, as no mention is made of it before A.D. 1175. It was commonly thought that Sumatra (Suvarnadvipa) and Java (Yavaidvipa) were separated by the bay which, in southern Sumatra, penetrated far inland. Only after 1175 Chinese and Arabic sailors found that Java minor (= southern Sumatra) and Java major (= the present Java) were separated by a strait.

It therefore appears that the geological evolution has been of great influence on the economical and political history of Indonesia.

Summary of the evolution of the Sunda region

The oldest traces of this evolution are of Devonian age. The process started with the subsidence of a narrow zone in the middle of the Indonesian Primeval Continent, the Anambas zone. From this central zone orogenesis migrated step by step in lateral direction, southerly as well as northerly. The orogenic system which migrated to the south developed very fully. During this southward migration the Jurassic Malaya Orogene, the Cretaceous Sumatra Orogene and the recent Sunda Orogene were formed successively. Each of these orogenes consisted (or consists) of a volcanic inner arc, a non-volcanic outer arc and a foredeep; each of them was situated several hundreds of kilometres more towards the present Indian Ocean than its predecessor (Fig. 33).

In northerly direction we find the Tertiary Semitau Orogene and the Quaternary Serawak Orogene. Both had a volcanic inner arc, a non-volcanic outer arc, and a foredeep. The migration here is directed towards the north; the Serawak Orogene lies more than 100 kilometres to the north of its predecessor.

At present a new continent of more than 800 kilometres diameter is situated between the Serawak Orogene to the north and the Sunda Orogene to the south. These young mountains at the borders of the Sunda Land have nevertheless a common ancestor. They are products of the orogenesis which started in the Anambas focus and which ploughed through the older crust, totally transforming and deforming it.

Other orogenic systems in Indonesia

The development of the Anambas system into the present configuration of the Sunda region has been described as an example of the orogenic evolution which takes place in Indonesia. A number of similar foci of diastrophism can be distinguished in the Indonesian Archipelago, whose crustal waves meet and overlap. Such a focus originated for example in Early Mesozoic times in the Macassar Straits region. From this centre, called Pulu Laut focus, a system developed which began to migrate in westerly direction. This is the Meratus Orogene, which runs at right angles to the Anambas system in western Borneo. From this same centre a system began to migrate towards the Pacific; this system is responsible for the structure of Celebes. Roughly speaking, the volcanic inner arc of this Celebes Orogene is formed by the northern peninsula, the western part of central Celebes and the southern peninsula; whereas the non-volcanic outer arc is formed by the eastern and south-eastern peninsulas. This Celebes Orogene turns its concave side outwards, in the same way as the Serawak Orogene. The foreland of the Celebes Orogene is formed by the Banggai Archipelago and the Sula Islands region, where older orogenic cycles had already formed a crystalline crust.

Another important focus is situated in the Banda basin. From this centre systems of crustal waves began to develop in Late Palaeozoic times; in northerly direction it was the Ceram system which began to migrate towards the Sula Islands and in southerly direction the Tanimbar system towards Australia.

Fig. 49 shows a diagram of the development of the Banda arcs. In broad lines it resembles the diagram of the Anambas system. There is, however, a remarkable difference between the two. In the Banda area the crust tends to sink and consequently the average level of the crust is about $1\frac{1}{2}$ kilometres lower than in the Sunda region.

After the orogenesis had ploughed through the Sunda area, the latter stiffened and formed a continent. In the Banda region the crust has been broken up and the older orogenic zones have foundered to great depths, giving birth to deep basins (Banda basin: 5266 metres; Weber deep: 7440 metres below sea level).

Apart from the above mentioned difference it is clear that the Banda System possesses the same structure as the Anambas System. The following structural elements may be distinguished: a central block (here developed as the central Banda basin), the volcanic inner

160

GEOLOGICAL EPOCHS	NORTHERN SECTION (CERAM SYSTEM)			OROGENIC FOCUS	SOUTHERN SECTION (TANIMBAR SYSTEM)		
	CERAM SEA	NORTHERN PART OF THE AMBON ZONE	NORTHERN OROGENIC BELT IN THE BANDA BASIN	CENTRAL BANDA BASIN	SOUTHERN OROGENIC BELT IN THE BANDA BASIN	SOUTHERN PART OF THE AMBON ZONE	TANIMBAR ZONE
YOUNG PALEOZOIC	SULA FORELAND			axis of the geo-synclinal sea			SAHUL FORELAND
PERMIAN			northern side deep	median volcanic uplift	southern sidedeep		
TRIASSIC		foredeep with Triassic Flysch	non-volcanic geanticline	Median basin	non-volcanic geanticline	foredeep with Triassic Flysch	
JURASSIC		foredeep with Jurassic flysch	volcanic geanticline		volcanic geanticline	foredeep with Jurassic Flysch	
CRETACEOUS	foredeep with northward overthrust structures		extinct-volcanic geanticline		extinct-volcanic geanticline	foredeep with southward overthrust structures	
PALEOGENE	foredeep with northward overthrusts	non-volcanic geanticline		block-faulted median basin	non-volcanic geanticline		foredeep with southward overthrusts
OLD NEOGENE	foredeep with northward overthrusts	volcanic geanticline			volcanic geanticline		foredeep with southward overthrusts
PLIO-PLEISTOCENE	non-volcanic outer arc (Ceram)	extinct-volcanic inner arc (Ambon)	extinct-volcanic geanticline	Central Banda Basin (Siboga Ridges)	extinct-volcanic geanticline	new volc. inner arc / extinct volcanic SW part of the Weber Deep	non-volcanic outer arc (Tanimbar)

PRESENT PROFILE

SULA SPUR	CERAM SEA	CERAM	AMBON	SIBOGA RIDGES	CENTRAL BANDA BASIN	NILA WEBER DEEP	BABAR SPUR	TANIMBAR	SAHUL SHELF
	Foredeep	Non-volcanic outer arc	Inter-deep	Extinct-volcanic inner arc		New volcanic inner arc (collapsed)	Extinct volcanic inner arc (collapsed)	Inter-deep	Non-volcanic outer arc	Foredeep

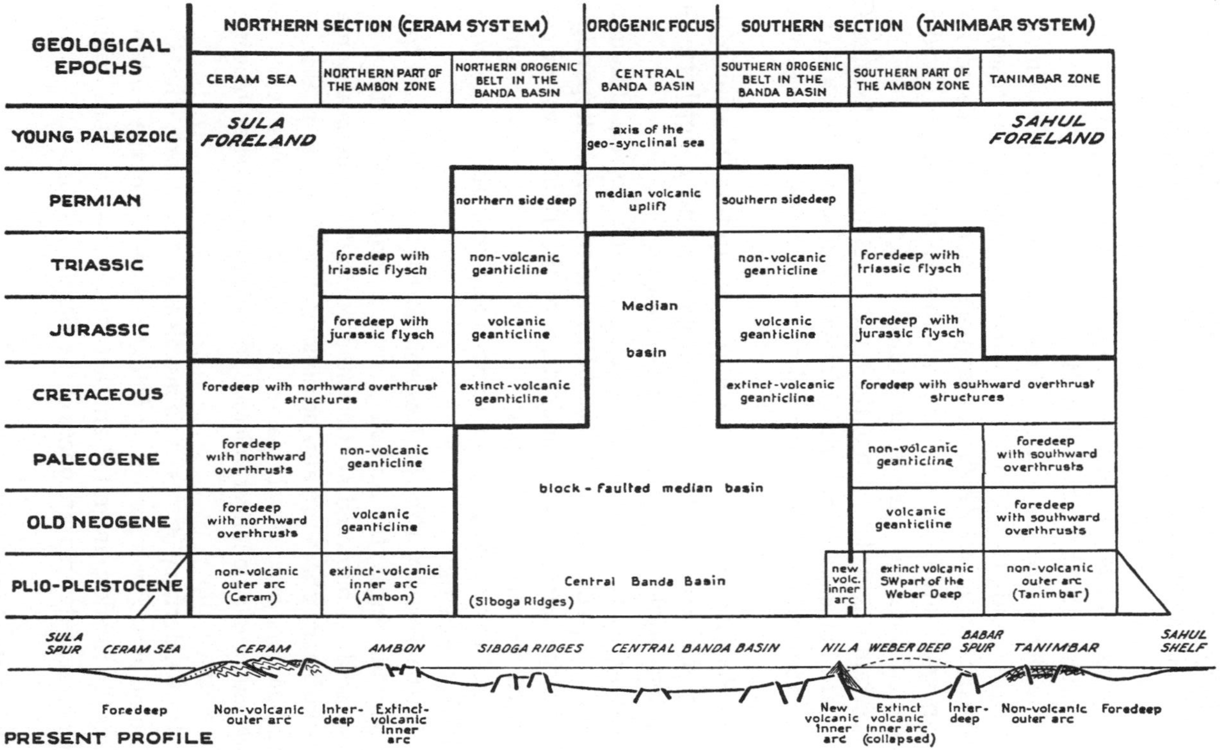

Fig. 49

arc (which runs from Wetar over Banda to Ambon), the interdeep (Weber deep), the non-volcanic outer arc with outward overthrusts and negative isostatic anomalies (running from Timor across the Tanimbar and Kai Islands to Ceram), the foredeep, and the foreland (Australia with the Sahul Shelf, Misoöl and Sula).

The Banda arcs are a classical example of their type of mountain systems. It appears that this regular structure is the result of an orogenic evolution which began more than two hundred million years ago (Late Palaeozoic) and which since then has evolved along identical lines, obeying identical rules.

Certain stages of growth of the Banda System can be distinguished. In Late Palaeozoic times the primeval continent in this area began to subside and eruptive material of the Atlantic suite was formed. This was the initial stage of the orogenic cycle.

In Permian times a ridge was pushed up from this initial basin which divided the focus of diastrophism into a northern and a southern border deep (the embryonic stage).

Through a series of uplifts, mountain chains developed from these border deeps during the Mesozoic Era, and the border deeps migrated farther outwards. This young stage of the Banda system can be divided into three sub-phases, namely a non-volcanic phase in Triassic, a volcanic phase in Jurassic and an extinct volcanic phase in Cretaceous times.

In the Tertiary Era the Mesozoic foredeeps were raised to form mountain ranges; the Ambon geanticline originated in this way and the foredeep migrated another step outwards. This nearly mature evolutionary phase can also be divided into three sub-phases, as the Ambon zone successively passed through a non-volcanic, an active volcanic and a no longer volcanic stage (in Palaeogene, Neogene and Quaternary times).

In Quaternary times a non-volcanic outer arc, the Ceram-Tanimbar arc, was pushed up from the foredeep of the Ambon zone and the foredeep migrated another step outwards. Thus originated the present-day mature phase of the Banda System.

Periodicity of orogenesis
One of the most important aspects of these cycles of orogenesis in Indonesia is the similarity of this process to a wave, spreading out from the focus of diastrophism. These waves must, however, not

be likened to the ripples on a pond after a stone has been thrown in. The lateral migration of the waves of orogenesis is jerky; during each jerk the axis of the foredeep moves another 100 to 200 kilometres outwards. Step by step, in arcs of increasing size, the orogeny migrates from the focus of diastrophism to the foreland. This process continues so long as the subsidence of the foredeeps can release sufficient energy for the formation of a mountain root, which in due course makes it possible for the foredeep to evolve into a new mountain chain. If this is no longer the case, the orogenic cycle comes to a stop and a period of continental stiffening of the crust begins (Sunda Land for example).

Mountain building appears to be a rhythmic process, of which the apparently quiet phases of subsidence are in fact the incubation periods for the following, relatively short phases of orogenesis. The stresses which accumulate in the depths during the incubation period, are relaxed during the orogenic revolution by tectogenic and volcanic processes in the earth's crust which are accompanied by earthquakes, gravity anomalies and other geophysical phenomena.

UMBGROVE called this alternation of evolutionary and revolutionary phases in the development of our planet "the pulse of the earth".

Processes of this type are called relaxation oscillations by the physicist. In this category belong widely different processes in nature such as the beating of the heart and the pulsating stars of the Cepheid type, as well as artificial processes, such as the ticking of a clock. In all these processes energy accumulates which is suddenly released. Most processes in nature which are associated with the dissipation of energy and where resistance plays a role, are periodic phenomena belonging to the relaxation oscillations. External conditions may influence the period, but they cannot alter the amplitude of the pulsation. The well known phenomenon of resonance is lacking. The interval, or time of relaxation (for orogenesis the time between two successive orogenic phases) can therefore vary between wide limits, but the intensity cannot exceed a given magnitude.

If we arrange the principal orogenic phases of the Sunda Mountain System in chronological order, we see that eight such phases can be distinguished with intervals of 20–40 million years. (fig. 50). The Banda System shows seven phases (no Carboniferous uplift is known) also with intervals of 20–40 million years.

The trends of the Indonesian mountain systems (fig. 51)

In chapter I (Physiography) we mentioned the fact that several great mountain systems meet in Indonesia; namely the Sunda System, the System of the East Asiatic island arcs and the Circum-Australian System. It appears from their geological history that these mountain systems spring from a number of foci of diastrophism.

The Sunda Mountain System (7,000 km long) which stretches from Assam to Banda, is composed of five mountain arcs belonging to an equal number of foci (Shan, Mergui, Anambas, Flores, Banda).

The eastern sectors of the Sunda Mountain System are interlaced with the northern and southern orogenic structures of the System of the East Asiatic island arcs. The foci of diastrophism in the Celebes deep and the Macassar deep (Pulu-Laut focus) belong to this latter system. From the Celebes-deep focus in a north-westerly direction the Sulu Archipelago orogen developed, in an easterly direction the Sangihe-Minahassa orogen and in a southerly direction the North Celebes orogen. The Meratus orogen migrated in a westerly direction from the Pulu-Laut focus and the orogen of central and southern Celebes in an easterly direction from this same focus.

Halmaheira, the Berau peninsula of New Guinea and the Northern Watershed Mountains to the east, all belong to a mountain system, which migrated in westerly and southerly directions from the submerged borderland of Northern Melanesia. This mountain system reaches as far as the Bismarck Archipelago and has a total length of approximately 3,000 km.

The Snow Mountains of New Guinea have been pushed up from the Papuan geosyncline, which borders the Australian continent to the north. This uplift has been preceded by a very long period of subsidence which started in Silurian times, about 350 million years ago. Comparatively little is known about the pre-Tertiary history of this geosyncline. The actual uplift of a median ridge seems to have begun in Oligocene times, but not until Plio-Pleistocene times were the border ranges pushed up to great heights. (Doorman Range to the north and Snow Mountains to the south). The central mountain. system of New Guinea is therefore still in a young stage of development.

General geological conclusions

We have now completed our sketch of the geological history of

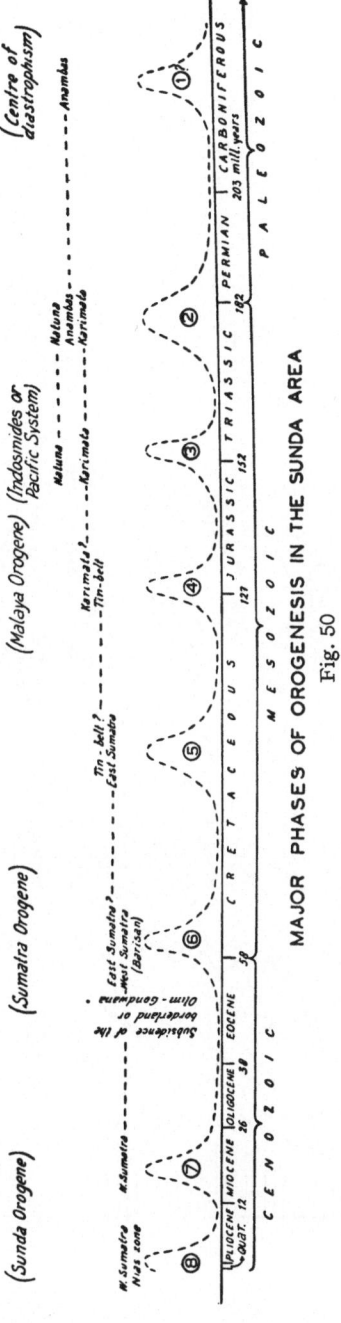

MAJOR PHASES OF OROGENESIS IN THE SUNDA AREA

Fig. 50

165

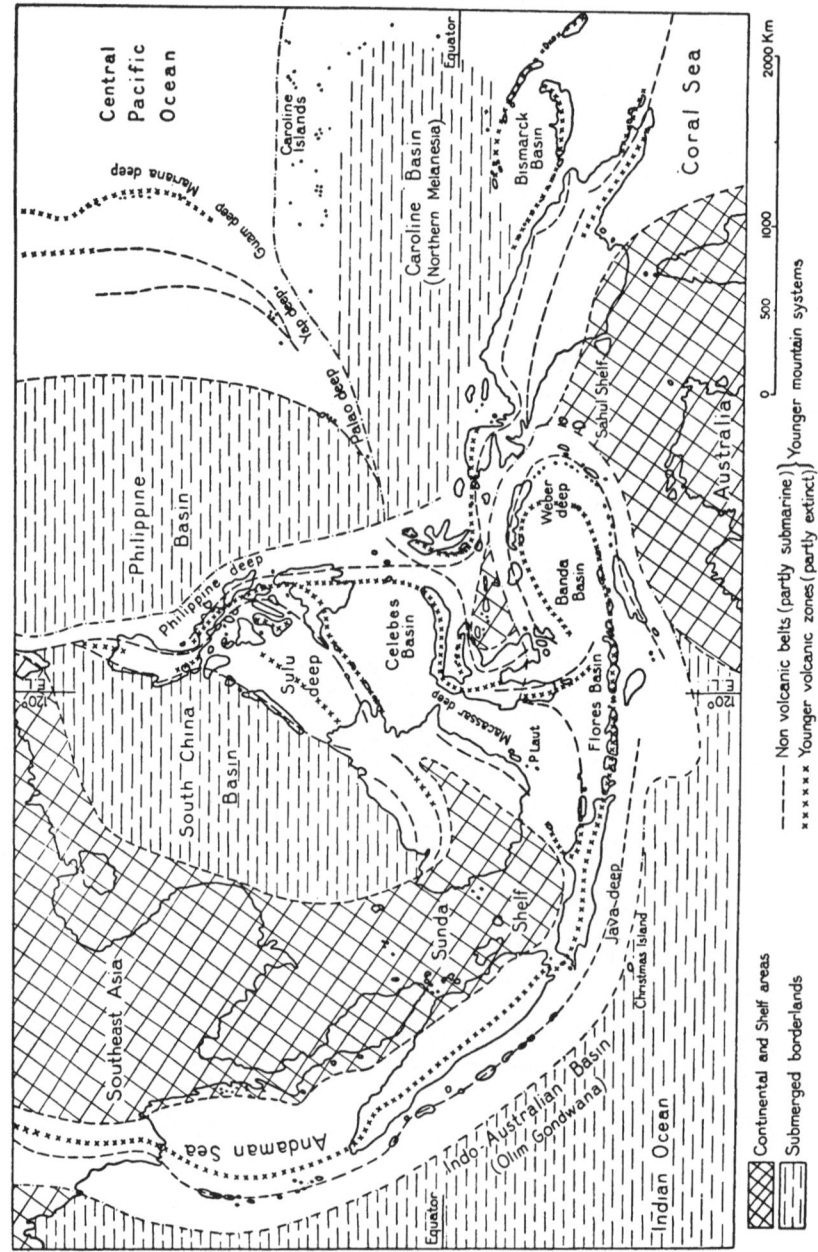

Fig. 51

Sketch map showing the Indonesian mountain systems.

Indonesia. What does this geological history teach us of the general principles of crustal evolution? A general characteristic is the periodical volcanic, tectonic and geophysical phenomena which change the structure and composition of the crust. There are constructive and destructive forces at work on the crust. In the very remote geological past a crystalline crust had been produced, which formed an extensive continental area. This Indonesian Primeval Continent was again partly destroyed in two different ways. In the first place, extensive areas began to subside and made way for the ocean; this is the epeirogenic foundering of borderlands such as Philippinia, Northern Melanesia and Olim-Gondwana. Secondly, orogenic cycles ploughed through the primeval continent and transformed and deformed the crust. This happened for example in the Sunda region and in the Moluccas. The final result of these orogenic processes can be either a regeneration of the crust, leading to the birth of a new continental area, as happened in the Sunda region of western Indonesia, or a final breaking up and subsidence of the crust, as happened in the Moluccas of eastern Indonesia.

The epeirogenic and the orogenic destruction of the primeval continent have therefore quite different results. Epeirogenic destruction leads to the formation of oceanic basins. This type of destruction seems to be of a permanent nature. In the other case the destruction may be followed by reconsolidation, as we observe in the Sunda region.

The evolution of our planet is a long chain of processes striving after equilibrium, therefore also after hydrostatic equilibrium. The latter is only possible when all relief of the crust has disappeared and the oceans surround the world as one continuous layer of about 3 km depth. As long as this state has not been reached, there will be stresses in the crust and a tendency towards gravitational flowage, causing land areas to disappear under sealevel. The fact that there is still land, is due to the enormous stores of energy inside the earth, which it inherited when it was born. This enabled the earth to produce again and again new mountains and continents during its life history of more than three thousand million years.

INDEX

Batang Toru valley 83
batholiths 68, 69, 76, 133, 136, 137, 148
batholithic intruston 86
bathyderm 95
bathydermal 5, 96
bathydermal spreading 142
bathydermis 110
Batu Basi 134
Baturadja 86
BEAUFORT, L. F. DE, 40
BELOUSOV, V. V., 16, 32
BELTZ, E. W., 146
bench-vice concept 14, 28
Benkulen 72
Berau 49
Berau peninsula 164
BERLAGE, H. P., 123
BERTRAND, M., 8, 33
bibliography East Indian Archipelago 40
bicausal-fixistic theories 16
bicausal interpretation 95, 109
bicausality concept 6, 18, 23, 25, 112
Billiton 73, 137
Bintan 135, 137
Bismarck Archipelago 43, 164
Blang Kedjeren basin 83
block diagram 128
BOEHM, G. 38
Bogor zone 109
border deep 162
borderland 42, 93, 144, 164, 167
Borelis (= Alveolina) 56
BORN, A. 8, 33
Borneo 39, 42, 46, 47, 48, 49, 62, 70, 73, 77, 106, 128, 133, 134, 137, 146, 147, 149, 150, 151, 156, 157, 160
Borobudur 101
BOWIE, W., 8
Brantas delta 152
BROUWER, H. A. 17, 38, 39, 40
Brunei 157
BRUYN, J. W. DE, 115, 156
BUCHER, W. H., 33
buckling down 29
buckling of granitic crust 150
Bukit Assam 60
Bukit Mapas 86
Bulan Straits 137
Bündner Schiefer 140
BURCK, H., 38
Burdigalian 58
Burma 115
Buru 38, 116
Buton 39, 116
Butung Sumpur valley 83

Cafemic constituents 22
Cainozoic 49, 51, 60, 76, 144, 145
caldera 88
caldera (Zandzee) 88
cambium layer 18
Cambrian 47
Carboniferous 46, 47, 141
Carboniferous flora 49
Carstensz peaks 98
Celebes 38, 39, 42, 43, 116, 117, 160
Celebes basin 117
Celebes deep 164
Celebes expedition 40
centres of diastrophism 15
Ceram 38, 53, 103, 116, 162
Ceram Sea 116
Ceram system 160
Ceram-Tanimbar arc 162
chain reactions 18
CHAMBERLIN, Th. C., 8
Charles Louis Mountains 43
chemical differentiation 19
chemical energy 115
chemical equilibrium 80
chemical processes 15, 80
chemical source of energy 16
chemical transformation 2
Cheribonian 55, 60
CHESTER LONGWELL 8
chloritization 22
Chinese districts 75
Christmas Island 64, 66, 145
Clathrodictyon cf. spatiosum Boehnke 48
Circum Australian system 42, 164
Circum Pacific system 42
CLOOS, H., 3, 4, 33
collapse 143, 155, 156
collapse structures 7, 90
collapse of volcanoes 98, 137
COLIJN, A. H., 40
comagmatic regions (provinces) 64, 72, 73
compression 6
compression phenomena 109
compressive force 150
compressive settling 96, 103, 106
consanguinic suites of igneous rocks 18, 64, 127
constructive forces 167
continental drift 7
continental growth 7
continental nuclei 8
contraction theory 5, 12
convection currents 6, 7, 9, 16
cooling of the earth 5, 9, 80
cope-stone blocks 92
corals 55
cosmogony 13

169

Insubric phase 143
interatomic forces 10
interdeep 75, 162
intermediate basaltic layer 18, 19, 122
internal energy 80, 115
internal equilibrium 80
internal heat 9
internal volcanism 62, 73
intervals of orogenic phases 72
intrusion 86, 128
Irian 43
irreversible chain reactions 15
island arcs (Circum-Pacific) 41
island arcs (East Asiatic) 164
isogams 115
isostasy 145
isostatic anomalies 23, 28, 29, 115, 153
isostatic anomalies (negative) 116, 117, 156, 162
isostatic anomalies (positive) 117
isostatic equilibrium 1, 23, 115
isostatic rise 6

Java 39, 64, 66, 77, 91, 93, 99, 101, 103, 105, 108, 109, 110, 111, 115, 116, 117, 119, 120, 123, 128, 139, 146, 147, 149, 150, 152, 155, 156, 158, 159
Java deep 51
Java major 159
Java minor 159
Java sea 75
JOLY, J., 6, 33
JONGH, A. C. DE, 39
JONGMANS, W. J., 135, 141
JONKER, H. G., 38
JUNGHUHN, F. 37
Jurassic 69, 71, 73, 135, 136, 137, 138, 159, 162
juvenile granitic magma 22
juvenile magmatic matter 24

Kabuh beds 156
Kai Islands 116, 162
Kalea formation 149
Kaliglagah 58
Karangkobar 101
Karimata zone 73, 75, 77, 78, 132, 133, 134, 136, 137, 138, 139, 142
Karimundjawa islands 64, 66
Katacycloclypeus 58
Kendeng ridge 117
Kendeng zone 152, 153
keratophyres 135
Kerintji 120
Ketapang-Matan district 75
KIENOW, S., 4

KLEIN, W. C., 39
KLOMPÉ, TH. H. F., 41
KOBER, L., 5, 10, 14, 33
KOENIGSWALD, G. H. R. VON, 58
KONING, L. P. G., 125, 126
KOOLHOVEN, W. C. B., 39, 146
KOPERBERG, M., 38
Krakatau 61, 88, 92, 110
KRAUS, E., 15, 16, 27, 28, 33
Kuching 49
KUENEN, PH. H., 40
KUHN, W., 12, 33
Kundur 135
Kutai 106

LACROIX, A. 12
LANDES, K. K., 5, 33
Lampong Districts 66
Lampong tuffs 155
landslides 98, 99
lateral-compression hypotheses IX
lateral displacement 6
lateral shift 12
LEES, G. M., 3, 33
Lembang depression 88
Lepidocyclina 58
LE ROUX 40
Lesser Sunda islands 40, 42, 62, 64, 91, 93, 109, 154
letter-classification of the Tertiary 55, 57
Lingga 135
Lingsing beds 139
Lombok 91, 109, 115
long-range migrations 14
Luzon arc 43
Lyell's principle of uniformitarianism 114

Macassar deep 117, 164
Macassar Sea 106
Macassar Straits 69, 106, 160
macrocosm 97
Madura Straits 117, 152, 153
magma 80, 110
magma (palingenic) 110
magmatic blister 12
magmatic processes 136
magmatic root 118
magma zone 110
magnetic field 114, 115
MAILLET, 4
Main Range 137
Maju 116
Malaya (n peninsula) 42, 49, 73, 77, 132, 137, 142, 158
Malintan 120
Mangani 145
Manondong 120

172

submarine K. XVIII 40
submarine ridge south of Java 66
sub-phases 162
substratum 6
subvolcanic rock 68
sucking down 29
suction theory 15
SUESS, ED., 5, 8, 12, 33
Sukadana 66
Sula islands 38, 160, 162
Sulawesi 43
Sulu archipelago 43, 164
Sulu basin 117
Sumatra 38, 39, 46, 47, 49, 62, 64,
    66, 71, 77, 86, 93, 103, 109, 110,
    111, 115, 116, 120, 123, 128, 134,
    138, 139, 142, 143, 145, 146, 149,
    150, 155, 156, 157, 158, 159
Sumatra-Java arc 154
Sumatra-Java geanticline 155
Sumba 116
Sunda continent 8
Sunda kratogene 129
Sunda Land 66, 71, 76, 78, 103, 117,
    119, 123, 125, 132, 163
Sunda region 51, 53, 73, 77, 78, 112,
    128, 129, 130, 134, 143, 144, 145,
    150, 154, 157, 159, 160, 167
Sunda shelf 42, 157
Sunda Straits 92, 110, 111, 155
Sunda Straits tumor 158
Sunda system 42, 164
Sunda volcanic complex 99
Surakarta region 109
Suvarnadvipa 159
system of East Asiatic island arcs
    42

Taeniopteris 49
Tambakan ridge 99
Tambang Sawah 148
Tambora 61
tangential compression 3, 5, 6, 86, 95,
    107, 109, 150
Tanimbar 116, 162
Tanimbar system 160
Tapanuli 120
Tatrot fauna 60
TAYLOR, F. B., 7, 33
tectogenesis 6
tectogenesis (gravitational) 92, 94,
    95, 96, 97, 98, 99
tectogenesis (primary) 6, 11, 14, 15,
    93, 94, 95, 98, 143
tectogenesis (secondary) 7, 11, 14,
    15, 94, 95, 109
tectonics 5, 93
tectonic denudation 4, 5, 95, 98, 99
tectonic relief 103, 106

tectonic slope 98
tectosphere 13, 18, 21, 121, 122, 125,
    145
Telen area (region) 46, 48, 76, 128
Telisa beds 149
Tello-Betic system 29
temperature 4
Tengger Mountains 88
tensile stresses 137, 145
tensional phenomena 109, 120
TERMIER, P., 12
Ternate 116
Tertiary 52, 53, 57, 71, 77, 143, 148,
    150, 156, 159, 162
Tethys sea 6, 19, 50, 153
thermal convection currents 9
thermal processes 15
thermal source of energy 16
thrusting 3
Tifore 116
Tigapulu Mountains 143
time-scale 96, 97
Timor 38, 49, 103, 116, 162
Tin granites 138
Tin range 137, 138, 139
Tin zone 73, 75, 78, 132, 137, 140
Tjandi hills 158
Tjidjulang 58
Tjikotok 148
Tjiodeng 55
Tjipitung 148
Toba cauldron 83, 85, 87
Toba (region) 72, 155, 156
TOBLER, A., 38, 39
tonalite 133
tooth paste (viscosity) 5
Tor si Hite 121
trachytes (alkali) 49
traineau écraseur 104
trajectories of maximum stress 95,
    103
transfer of energy 80, 93
transformation 16, 44
trans-uranium elements 10
trends of mountain systems 164
Triassic 73, 77, 132, 133, 134, 135,
    136, 137, 162
Trillina howchini 58
Trinjl 38, 58
Tukang Besi Islands 116
tumor 92, 110
tumor (Batak) 85, 86
tumor (Gedongsurian) 86

Ularbulu ridge 157
ultra-metamorphism 22
UMBGROVE, J. H. F., 5, 15, 16, 17,
    33, 54, 55, 150, 163
Umbilin 146

176